Butte's Memory Book

BUTTE'S MEMORY BOOK

by
DON JAMES

Pictures by
C. OWEN SMITHERS, SR.

THE CAXTON PRINTERS, LTD.
CALDWELL, IDAHO 83605
1980

First printing December, 1975
Second printing November, 1976
Third printing January, 1980

Library of Congress Cataloging in Publication Data

James, Don.
  Butte's memory book.

  Includes index.
  1. Butte, Mont. — History. 2. Copper mines and
mining — Montana — Butte. I. Smithers, C. Owen.
II. Title.
F739.B8J35   978.6′68   75-12297
ISBN 0-87004-245-9

Lithographed and Bound in the United States of America by
The Caxton Printers, Ltd.
Caldwell, Idaho 83605
132961

For Connie.

CONTENTS

PREFACE

In a book's preface the author customarily explains why or how it was written. Also the preface frequently includes pertinent acknowledgements, if not too many, in which case they receive a page or so for themselves.

This whole task is somewhat difficult with our book, because, in essence, it is *our* book. The substance consists largely of pictures from the Smithers historical collection, owned by C. Owen Smithers, Sr. The idea for the book was nurtured by The Caxton Printers, Ltd., who decided to publish it. Final selection of contents, writing, editing, and much of the rough layout for the book fell to me.

For years people have said, "Someone should do the *real* Butte story." Actually many books about Butte have been done. Some deal with the less inhibited and sometimes violent old Butte. Some dramatically describe the war of the copper kings. Novels have been backgrounded against the Montana city sprawled upon the "richest hill on earth." Although these books and many shorter pieces contribute to the history, some critics still lament that none has caught that elusive "real" Butte story.

If the story has escaped a true and complete telling, possibly too much attention has been concerned specifically with the mining camp itself, the companies, the unions, some of the people, or, of course, the "action." Possibly there is too much for a single writer, or for one book. The history may read so much like fiction that the historical story defeats itself in the telling. Nor does this book by any means tell the complete story. We do hope, however, that it presents a different story in a different way because there *is* a Butte story — real, honest, and exciting.

The book came about because C. Owen Smithers, Sr. for many years one of Montana's most prestigious professional photographers, a native of Kalispell and a long time resident of Butte, had a dedicated and successful zeal to collect a photographic history of the state and of Butte. In the 1960s, several years before his death in 1973, he interested Jack Murray, a representative of the outstanding regional publishers, The Caxton Printers, Ltd., in the possibilities of publishing some of the pictures.

Gordon Gipson, whose family founded the publishing firm, became enthusiastic about a Butte book that would include suitable text as well as a Butte selection from the Smithers photographic collection. Caxton already was publishing a reprint of the turn-of-the-century Harry C. Freeman book, *A Brief History of Butte, Montana,* and Mr. Gipson believed that subsequent Butte history was highly significant to regional chronicles. The Smithers collection offered a unique opportunity to present some of it.

In 1968 Mr. Murray and Mr. Gipson asked me if I would be interested in selecting the pictures, organizing the format, and writing the text and captions for the book. I had been reared in Butte and had strong roots there. Mrs. James had been Connie Belle Hoover of the Butte Camp Fire Girls. At the time I was under contract for another book, but I was interested in a Butte book and agreed to consider doing it.

Shortly after my conversations with the Caxton people Mrs. James and I visited Butte where Owen ("Smiggs") Smithers and I had our last visit. We had been friends since the twenties and we were in rapport about the possibilities of a Butte book. Upon my return to our home in Portland, Oregon, I became seriously ill for many months. The proposed project lay dormant. When I resumed work, Mr. Murray brought me about 900 photographs that Owen Smithers had sent to Caxton. Some pictures were captioned, along with a relatively small amount of commentary in rough notes.

After studying the material I began to visualize a book combining lean text and captions with the historical pictures to form a nostalgic word-picture portrait of Butte that many persons would recognize in part or in whole. I discussed the idea with the Caxton people. They and Smiggs agreed with the plan.

Before the contract was signed, C. Owen Smithers, Sr. died in February, 1973, a few months before he would have been 80 years of age. The decision was made to continue with the book, relying upon help from C. Owen Smithers, Jr., who had taken over the full management of the photographic firm.

So — with considerable effort, enthusiasm, selection, time, frustrations, enjoyment, and some sadness — we have put together this Butte book. It is not a minutely documented history. The text is not particularly erudite because essentially it was inspired by many persons who have had a lifelong and sometimes slightly emotional love affair with Butte. It is not a sociological, economical, or political survey. To those who seek such studies in detail, or "in depth" as researchers like to say, we refer you to the various publications about Butte which may be found in most sizeable libraries. This is simply a picture and word book about the great mining camp and its people.

Today we hear talk that Butte is finished — that the city verges on becoming a ghost town, that no one is left, that it is a poor unrelated shell of its former self. Certainly the former resident who returns after a considerable absence will find many changes. Decadence permeates some older neighborhoods. Streets are caved. Downtown fires have taken a heavy toll. On the hill, old houses appear to be woefully dilapidated, unpainted, and sagging. All in all, a good deal of thinking says that "something is the trouble with Butte."

"The trouble with Butte," according to an old-time newspaperman, "is that it had a sugar daddy for too long. He kept old buddies on the payroll when they should have been laid off, and mines going when they should have been shut down. He even fed miners striking against him."

Another older resident says, "The trouble with Butte is the old-timers are almost gone. Newcomers moved in and they don't understand or care."

"There's no big payroll," says a businessman. "Pit mining doesn't need many miners. They're shovel operators, drillers, and truck drivers. A real, honest-to-God hardrock miner is getting hard to find."

An old hardrock miner says, "The trouble with Butte is they closed it up. When it was a hell-roaring, wide-open, hard-drinking, free-gambling camp we were better off."

Actually very little may be "the trouble with Butte" and people may be observing mostly the effects of progress and aging. After World War II, The Anaconda Company spent about $80 million developing new methods to economically mine vast quantities of ore still available around Butte. The resulting lower employment demands and the visible ecological effects doubtlessly are dramatic to residents, but mining continues to be the major industry, including important underground as well as pit operations.

Other indications of survival are evident. Not the least of these is, in essence, a long extension of the West Coast seaport of Seattle to bring the mountain city more industry and payrolls as a distribution center. Some remaining hardrock miners may be dismayed that one of the world's great mining camps should become part of a seaport on an ocean 600 miles and two states away. Other residents may smile enthusiastically and say, "It's progress."

All this does not change the fact that the old Butte did exist. Its history is fabulous. Its people have been famous. If its future should be unpredictable, no one who lived in Butte in the old days — and some of the newer days — can ever forget the great copper camp. The memories are treasured by thousands who knew Butte well.

We bluntly admit that this book is steeped in nostalgia and deliberately planned to touch the hearts of many generations who made Butte their home. If you have the capacity to dwell in memories, to remember how it was, to recall old places and old faces, to explore the past, and possibly to manage a wet eye now and then, we welcome you. In a sense this is Butte's "memory book."

We just thought that someone had better do it before the old Butte disappears into a great pit.

ACKNOWLEDGEMENTS

Acknowledging those who in one way or another contributed to putting together this book would be an almost endless duty, and would, of necessity, go back over many years to many persons who each contributed in one way or another.

As the book began to take shape, certain persons contributed contemporarily and materially and must be mentioned at once. In Butte they include: Mrs. Tana Keith, Mrs. Ann Lynch, Tom Wigal, Charles Goddard, Frank Quinn, Carl Rowan; and Joyce Bouchard and Eileen Riley of the Butte library staff. In Portland copious thanks are due former Butte residents Larry (Lala) Manion, the Don Arel and Hosty families, and Thomas P. Driscoll for their aid. Special thanks are due Mrs. Florence Brentano for her aid in manuscript preparation.

Nor can I overlook some who in years past substantially contributed to my own feeling about Butte that I have tried to put into this book: Law Riskin, George McVey, Jean Jordan, Sam Parker, Led Stormes, Charles and Violet Darlington, Mert and Frances Callow, and my own James, McDonald, and Lilly relatives who lived in Butte. There are far too many others to find room for mention, including alumni members of my Butte High School class of 1923!

Finally, I must especially thank C. Owen Smithers, Jr. for his aid in continuing his father's work, and publisher Gordon Gipson, and Jack Murray who represents the firm, The Caxton Printers, Ltd.

DON JAMES

INTRODUCTION

C. Owen Smithers, Sr. might well be described as the "Photographer Historian" of Butte and of the Treasure State.

A native of Kalispell, Montana, he was a veteran of military service on the Mexican Border and in World Wars I and II. He came to Butte in 1921 as a news photographer and founded his photographic business shortly afterwards. In the intervening years his studio became, and still is, a focal point for writers from all parts of the United States and many foreign countries. They sought and received historical prints to illustrate writings in their particular fields.

As a longtime Butte newspaper reporter, I know of no individual to whom I could pay a tribute that would be more worthily deserved than my friend, known to the newspaper fraternity as "Smiggs," "Flash-Bulb," and "Smithers."

I am one of those who had the privilege and good fortune to come under the wise, conscientious guidance of "Smiggs" when I was a young reporter covering assignments on which he did the photographic work.

We of the newspaper profession in those years, and later, watched his career with great pride and were especially pleased to know that the great majority of the public of Butte, and Montana, in general, were aware of Mr. Smithers' contributions to the development of his community and state through his photographic skill and historical knowledge.

FRANK P. QUINN

Butte's Memory Book

THE MINES

The mines of Butte always have been the true essence of Butte. Probably no one who has lived there for any considerable time has not been attuned to the mines and usually dependent upon them in one way or another. And as the people of Butte created the unique lifestyle of the great mining camp, so did the mines create the lifestyle of the people.

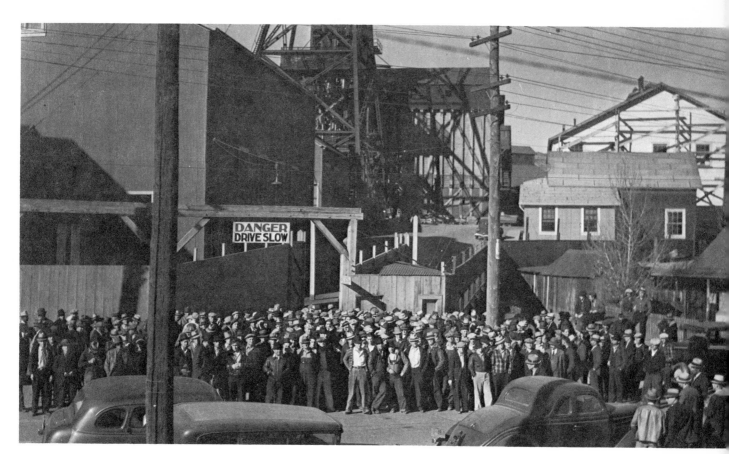

Rustling for jobs at the gate of the Leonard during the Great Depression.

THE MINES OF BUTTE

The ore was there through the ages before 1856 when Judge Irvine and his explorers found the old "diggings" and elkhorns used as gads. Doubtlessly they were curious, but they traveled on, not knowing that tremendous wealth lay beneath surrounding surface land.

About that time rumors of gold already had been heard elsewhere in Montana mountains. In 1858 James and Granville Stuart began gold mining at Gold Creek 20 miles west of Deer Lodge. Other Montana gold strikes followed and the names are familiar in the state's history: Bannack, Alder Gulch, Virginia City, Confederate and Last Chance gulches. Then in 1864, G. O. Humphrey and William Allison discovered gold on Butte Hill.

Historians generally recognize three important mining eras for the camp. First came the gold era that lasted about ten years. When the gold was about worked out, miners found silver and Butte prospered until the silver crash and panic in 1893.

Meanwhile, copper had been discovered in the mines. Eastern capital was flowing into town and after 1900 Butte boomed through World War I until the Great Depression. In 1919, when the city's population exceeded 100,000, mining employed more than 20,000 workers.

Eventually only two major companies remained and the W. A. Clark interests finally sold out to the Anaconda Company on August 22, 1928. One by one famous mining shafts became idle as the result of consolidations and new methods of mining. Then in the 1950s open pit mining came to Butte and "the richest hill on earth" with its towering headframes and surface works began to disappear.

Geologists estimate that at least half a billion tons of ore remain to be taken from the area, and mining may go on for as many years as they can find ways to mine the reserve. Most of the mines pictured here are gone forever.

Headframes stand solidly over mine shafts. They are topped by sheaves, grooved wheels over which run heavy cables to raise and lower cages and skips in the shafts. For years most Butte people called them gallows or gallus frames. Kids and adults well knew the sounds of hoisting and roar of ore dumped from skips into huge bins.

4

Most mines had almost primitive beginnings: windlass and bucket for waste and then — hopefully — ore. Supposedly this is the first prospect shaft for the Pittsmont mine.

Marcus Daly bought the Anaconda mine in the 1880s and built an impressive hoisting plant. This was the establishment of what eventually became The Anaconda Company.

ANACONDA HOISTING WORKS—Butte.

B.& M. Smelter. North Works.

B.& M. Smelter. South Works.

Moose

Tunnel.

BOSTON & MONTANA Cº

Mountain View.

At first were the mines. Then came the smelters. The Boston and Montana Company properties before 1900 included some of Butte's most valuable holdings. In 1899 the company was absorbed into the massive Amalgamated Copper Company.

Generations of people grew up to accept these massive structures as part of the Butte silhouette. Each one reared high over a shaft, and each shaft had its own history of owners, miners, wealth, failures, exultations, and tragedies. Some names have been all but lost. Some made international history.

Alice Mill - Walkerville - 1880

ALICE MILL

In 1876 Marcus Daly bought the Alice silver mine for the Walkers, his Utah employers. It was Daly's first Butte operation. He built this mill to process Alice ore. In 1880 he sold his interest and bought the Anaconda. Walkerville was named after the Walkers.

PARROT SMELTER

Butte had three Parrot mines, one said to be the camp's first commercial mine. Smelters were built in 1867, but were not successful. Later, open hearth smelters were shut down by outraged citizens suffering from smelter fumes. This early Parrot smelter was at the southeast edge of town.

MOUNTAIN CON

Originally named Mountain Consolidated, miners found it easier to call it the Mountain Con. Here many an Irish lad worked for the colorful mine foreman, Jim Brennan.

Mt. Con Copper Mine
Butte, Mont.

The Minnie Healy, a Heinze mine, where his miners and those from an adjoining Amalgamated mine reportedly fought with guns, dynamite, boiling water, and ammonia.

10

MOP SMELTER

The famous Montana Ore Purchasing smelter that young Heinze built in Meaderville with eastern backing. When the smelter was in operation, Heinze began to lease and buy mines. He was embarked upon his startling and dramatic career.

CLARK'S COLUSA SMELTER

By 1885 W. A. Clark owned or was part owner of 46 mines. It was natural for him to extend his mining operations to reduction works and smelter. The Colusa smelter was only one of his many properties.

Smithers
Butte

THE SEVEN STACKS OF THE NEVERSWEAT

Before the advent of electrical hoists, the Neversweat mine probably was the most photographed and famous mine in Butte. Although it stood high above Irish-named Dublin Gulch on the west side of the hill, it was a favorite mine for Cornish "Cousin Jacks." One legend asserts that many a Cousin Jack was "ticketed in England with a tag reading 'The Seven Stacks of the Neversweat' and that Ellis Island employees would know where to send the new settlers."

The mine won its name because in its early development it was unusually cool and comfortable for workers. As the mine deepened, this advantage gradually disappeared. So did the famous stacks when electricity took over. Because it was close to town, spectacular, and easily accessible to amateur photographers, Owen Smithers once remarked, "Thousands of pictures of the Neversweat must have been snapped!"

None could be more spectacular than this one from the Smithers Historical Collection.

Timber Butte Mill sprawled down the hill south of the city. W. A. Clark built it to treat ore from his zinc mines. Anaconda bought it when Clark sold out and eventually tore it down.

The Black Rock mine, once a Butte & Superior Mining Co. property, was one of the world's greatest zinc producers.

Cabins in the shadows of the mines

The sounds of mining at their doors

As miners found work in the new mines, they built homes near their work. The clusters of homes became small settlements such as Walkerville, Centerville, and Meaderville. In time they spread out and blended to become part of expanding Butte. As a result, a large residential perimeter around the big hill lay within the sights, smells, and sound of the mines. This picture shows part of Meaderville at the eastern base of the hill.

And the mines became a living part of Christmas

Smithers Butte

Anselmo Mine Hoist Butte Montana-

THE MULES
"My sweetheart's a mule in the mine.
I drive with only one line.
On the dashboard I sit
And tobacco I spit,
All over my sweetheart's behind."
(An old miner's ballad.)

18

St. Lawrence 1898

Jim Ledford, a miner who lived in a cabin below the Anaconda mine, noticed that mine water running through discarded cans and iron junk made the metal disappear and left a heavy sludge. He was curious and had the sludge assayed. It was 98 per cent pure copper. He got a one year contract for water flowing from the Anaconda and earned $90,000. The company refused to renew the contract. They called it "precipitating" and this was the first plant. Today they refer to it as "leaching."

ST. LAWRENCE MINE 1896

Old man with lunch bucket

Mountains and hills around Butte were
well timbered. Mines needed tremendous amounts of stulls and braces. Smelters needed cordwood. Logging for mining depleted the timber. Smelter fumes
finished the job.

In later years demands became heavy on distant forests as more than 10,000 miles of mine workings were driven underground.

Timbering

Drilling

Tamping a round to blast

Machines speed work. Once miners shoveled high grade into cars. The shoveler was a mucker and his shovel was a muckstick, often cut short of handle for mucking in tight places. Light came from candles and later carbide lamps. Today machines do the mucking. Batteries power lights. But the ore is as high grade as ever.

The mules are long gone . . .

. . . and it's a long way to the surface

In the summer the going wasn't so tough. Big horses held back wagonloads of ore going down the hill from mines to the smelters. In the winter snows and spring mud it got rough, and trips up the hill with machinery and supplies sometimes called for ten or more horses.

Some of the best drivers in the West handled the work wagons of Butte, reins in skilled hands, voices in command, bullwhips ready, and heavy feet on brakes for the downhill trips with tons of wet ore.

Eventually the first electric ore trains appeared on Butte's streets. Meanwhile, the Anaconda Company built its own railway to the Anaconda smelter and called it the Butte, Anaconda & Pacific Railway. It initially was steam powered but was electrified in 1912. Clark hauled railroad cars of ore to his Timber Butte mill over his passenger streetcar rails later on.

On a bitterly cold night the scream of ore train car wheels against cold railroad track on 'the hill' could be heard all over town.

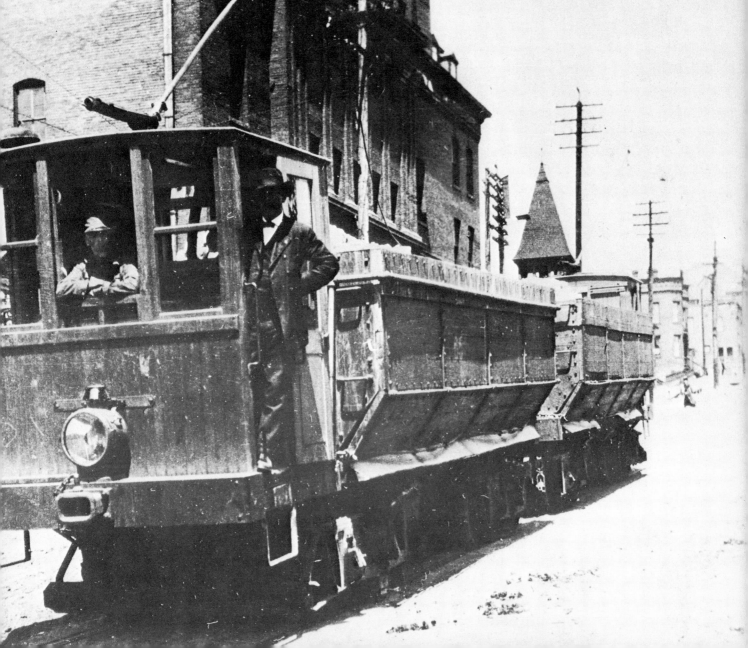

Not all mining operations in Montana
have been big operations. Sometimes the
prospector's gold pan has been important.
This is the way it began.

During the Great Depression more than
one miner discovered that it paid off to
pan for gold in Missoula Gulch which ran
through Butte's west side.

Champion Mucker
of the World
1200# in 1 minute - 44 seconds

MINER'S FIELD DAY

For years Columbia Gardens was the scene of Butte's annual Miner's Field Day. Before the age of mechanization, mucking and drilling contests offered high points of interest, as well as good gambling possibilities.

Butte claimed world champions in both hard rock skills. One driller, Walter Bradshaw, is said to have won $35,000 in prize money around the world. Mucking contests could be very close. In 1930 Inar Norgaaz, of the Badger mine, defeated the reigning champion, John Espelund, by one second when Espelund's shovel hit a snag and he had to hoist one more shovelful than the winner.

In later years, the contest among first aid teams probably was the main event. Team rivalry ran high, and the expertise of Butte teams attracted national recognition.

Hopefully the weather would be good for the day so that families could enjoy picnics up in the groves. And many of those families trooped to the grandstand to watch their menfolk perform as drillers, muckers, or a skilled first aid man on a team.

Butte has been a union town since June
13, 1878 when the Butte Workingman's
Union was organized with 261 members.
Union history from that date on has been
marked by constant activity, changes, and
occasionally intense friction. There have
been strikes, violence, solidarity, and
dynamitings. Whatever the changes in
union alignments may have been, one fact
remains clear: Butte has been a union
town!

A comparatively recent dispute between
two unions almost terminated in a riot, but
was averted when one union surrendered
the union hall to these angry members of
another union.

AND ONCE THEY BLEW UP THE OLD MINERS' HALL

The story might as well be told chronologically. In the first week of June, 1914, miners in the American Federation of Labor withdrew from the union in protest against heavy assessments for benefits to striking miners in Michigan. They organized an Independent Mine Workers' Union with about 4,000 members. A miner from the Speculator mine, Muckey McDonald, eventually was elected president of the new union.

On June 12 miners attacked several Miners' Union representatives at the gate of the "Spec." The next day disgruntled miners attacked and disrupted a Miners' Union Day parade. After the melee they stormed the Union Hall on North Main and demolished furnishings and fixtures. In the evening they took a safe containing Miners' Union money and records, loaded it on a van, took it south of town, and dynamited it open.

On June 20, Charles Moyer, Western Federation of Miners president, arrived in Butte. A few days later, on June 23, about 100 of Moyer's men opened rifle fire from Union Hall upon thousands of miners gathered around the building. The miners returned the fire and Moyer fled with his men. The attacking miners took cases of dynamite from nearby mines and blew up the hall.

Before peace was restored Mayor Lewis J. Duncan shot and killed a frenzied, knife-wielding miner in self-defense. The ACM rustling card office at the Parrot mine was blown up. Martial law was declared in Butte by Governor Sam V. Stewart. Muckey McDonald fled. Open shop was imposed on all Butte mines. The sheriff and mayor were deposed from office. Finally on November 5 martial law was ended and the camp settled down to the mildest, warmest winter in its history.

Some said it was high time!

IN MEMORY OF THE UNIDENTIFIED DEAD
OF THE GRANITE MOUNTAIN FIRE
JUNE, 8, 1917

FRANCIS H. ALLEN	PAUL KRELICH	WILJAM PALO
NILS ANDERSON	PETER KELLY	TONI PRETTI
HENRY BENNETTS	TOYVO KOKKONEN	CLAUDE POWERS
JOHN B. BIXBY	DAN KENNEDY	GUST POSHORSKY
JOHN CAVALLA	BEN KONENCY	JIM PONDERISKIS
JAMES CAVALLA	MIKE KUBILUS	FRED PAINTER
TIM J. CALLAHAN	MARION J. LIVAICH	TOIVO RINTILLA
ERNEST L. CHAPMAN	MARTIN LIVAICH	JOHN SEMINKIN
JOHN CHYTIL	EMIL LEHTI	THOS. SMITH
TOM DILLON	BAT LEARY	SALLE SULEMI
PAT DOHERTY	JOHN P. LALANNE	FRANK SKONDULA
EMIL ERICKSON	ANDREW MIKANIKOFF	GUST SALLIS
ALEX. FRANK	LAWRENCE MURPHY	MELCHIOR SHELDRUP
PAT FITZSIMMONS	MIKE MURRAY	FRANK SEMICH
MIKE GALLAGHER	NICK MAJESKY	KENNETH TOMBE
ERIK GRANBECK	GEORGE MILLICK	VERNON THOMPSON
EBEN GUSTAFSON	ARTHUR H. MURRAY	OLIVER J. WILLIAMS
PETER GALFI	WALTER MORRIS	NILOR WILSON
MIKE GRETHAS	N. R. McDONALD	SAM D. WUITUK
TOM HARRIS	JAS. A. NEGLEY	CHAS. H. WELCH
PETER HASTINGS	JOE A. NOVAO	NICK XELROS
TOM L. HUFF		MIKE YUKONOVICH

ERECTED BY NORTH BUTTE MINING Cº

DISASTERS IN THE MINES

The history of mining is heavily accented by disaster. All men who mine the earth for minerals accept the challenge of falling rock, fire, deadly gas, injuries, and death.

The mines of Butte have offered no exceptions. Their history is marked by disaster, sometimes massive in scope, and tragic in consequence. Nothing could strike more terror to the people than the ascending wail of the camp's many mine whistles as one by one they joined into a cacophonous alarm that announced disaster in a mine. Terrified women whose men worked underground grabbed children and ran toward the mines. Anxious crowds gathered at the gates. Emergency vehicles rushed toward the big hill. Newsboys raced through streets with hastily printed extras.

Finally details became known, bodies were brought up, names listed. Miners became heroes. Some died trying to save others. Death, mourning, and burial hung heavy over the city, and once, after the Granite Mountain and Speculator fire, 166 miners were buried in a mass funeral. Those who could not be identified are commemorated in bronze at the cemetery site.

November 23, 1889 marked Butte's first major mine calamity when fire in the Anaconda killed six. In 1893 a fire in the Silver Bow killed nine. On April 24, 1911 the Leonard hoist ran out of control and dropped 14 men 1500 feet to the sump. Five died. On September 3 in that year six young Black Rock nippers were killed in a hoisting accident.

Sixteen mine bosses stood around the Granite Mountain shaft, October 15, 1915, waiting for the after-lunch whistle signal to go below. Twelve cases of dynamite waited beside them. The big whistle sounded. The dynamite exploded. No one knows why. Parts of the 16 bodies were found as far as a mile away. On February 14, 1916, fire on the 1200-foot level of the Pennsylvania asphyxiated 21 men.

In 1917 the U.S. entered World War I. Mines operated around the clock. On the evening of June 8 hundreds of men had been lowered into the adjoining Speculator and Granite Mountain properties and were hard at work.

Earlier that day workers had been lowering a heavy cable down a Speculator air shaft. It had become fouled. Now an assistant foreman inspected the frayed insulated cable. His carbide lamp accidentally brushed the cable. In moments, aided by a strong up-draft, the entire 3,000 feet of timbered shaft became a raging furnace. Deadly smoke and gas crept back through mine tunnels.

Some miners escaped into adjoining mines. Gas trapped others. Eventually helmeted rescue crews could begin the search for the living or the dead. They found the dead — 166.

On June 10, before the search, while those on the surface waited for the fire to burn out, a signal suddenly flashed in the Speculator hoisting house. A startled engineer sprang to controls of the huge engine. A cage dropped to the indicated level far below. A brief wait. A signal to hoist. The cage surfaced and nine exhausted men stepped off. Sixteen others waited below. A young miner, Manus Duggan, had supervised construction of a bulkhead against gas, but he had died with three followers in an attempt to find a way out.

Men have always found death in Butte mines and probably always will as long as the mines are operated. But men who work in the mines accept the challenge.

Smoke coming from Granite Mt. Shaft fire morning after fire

June 1911

FIRE AT SPECULATOR MINE

A mine fire seldom is spectacular on the surface, but smoke billowing from a mine shaft strikes terror to those who wait above while men die below. This picture, with its penned side caption, graphically tells a portion of the tragic story of the Speculator and Granite Mountain fire. Two years before, 16 men were killed at the collar of this shaft when stacked boxes of powder exploded as the men waited to be lowered to work in the mine.

The Silver Bow mine rescue crew, above right, was a forerunner of highly trained "first aid" teams recruited from employees in most Butte mining facilities of any size. Some are in action here, right, during a miners' field day contest. They worked with split-second precision.

L. Coutts, J. J. Williams, L. Pope, L. Mundy, Callahan, F. Pope
This rescue crew worked heroically in the 1893 Silver Bow mine fire that killed nine men.

THE COPPER KINGS

No history of Butte could be significant without reference to the "copper kings." These were the men who discovered the potential of the mines, and had the vision, capabilities, and leadership to make all of it become reality.

In the beginning a good number of names surfaced in the camp's chronicle. Some have joined the legion of lost names in our passing history. Some remain of considerable importance for those who search out Butte's genesis.

Few may now remember G. O. Humphrey, William Allison, Dennis Leary, H. H. Porter, Tom O'Connor, or George Newkirk. They began to mine in Butte and inspired the first townsite. Leary and Porter built a smelter which wouldn't work. So did Joseph Ramsdell and William Parks. Newkirk located the famous Parrot Lode. W. L. Farlin struck it rich. Michael Hickey opened a mine in 1882 and called it the Anaconda. Many others prospected, staked out claims, and might have survived as one of the "greats." Few did.

Eventually, three men emerged as the "copper kings": William Andrews Clark, Marcus Daly, and F. Augustus Heinze. Other mine owners and operators existed for a time, but eventually the important properties fell to these three in one manner or another. Then Heinze sold out and two major companies remained. Finally, the Clark interests sold. Only one company survived, and that one gradually ceased to be identified with only one or two men in its steady growth into a mammoth, international organization of many interests. The old-time copper kings of Butte were gone.

THE WAR OF THE COPPER KINGS

Articles and several books have been devoted wholly or in part to Butte's historic "war of the copper kings" that actually began in the 1870s and reached a climax in the early 1900s. The battle of men, miners, and moguls involved courtroom scenes where some of the highest paid lawyers in the country fought long and tedious forays. It also included physical violence and traumatic warfare on the surface and in the depths of the mines.

Although the struggle eventually engaged Standard Oil and the Rockefellers, the Boston copper barons, and no small amount of Montana political connivance, most pertinent action revolved around William Andrews Clark, Marcus Daly, and Frederick Augustus Heinze.

Rights to rich minerals became a long and somewhat scandalous battle of litigation in Montana courts. Control of ore veins, by an 1872 Federal law, was determined by ownership of land on which veins apexed or surfaced. Heinze used his lawyers and the law so successfully that eventually he defeated the Standard Oil crowd, who, with Daly, had organized the Amalgamated Copper Company. They bought him out of Butte with ten and a half million dollars, which he later lost in Wall Street.

Marcus Daly died before the war ended and John D. Ryan most successfully inherited his leadership, along with Cornelius Francis Kelley. W. A. Clark survived very well until his death in 1925. Three years later his interests were sold to Anaconda Copper. Amalgamated, as is the way with the complexities of corporations, conglomerates, and such, eventually became Anaconda Mining Company and finally just The Anaconda Company.

WILLIAM ANDREWS CLARK

William Andrews Clark, a slight, precise man of Irish descent, born in Pennsylvania in 1839, was a school teacher in Missouri, and came west in 1863. He worked a mining claim in Colorado, then journeyed to the new Bannack gold strike in Montana and established some claims.

He quickly decided that he might more certainly make a fortune as a merchant and businessman. He brought food, tobacco, and other scarce supplies to Bannack and Virginia City and sold at high prices. He prospered, moved to Deer Lodge, and became a banker along with his merchandising. Soon he was the area's leading businessman.

In 1872 he went to Butte where placer mining had dwindled out. Only a few miners remained. They eagerly showed Clark their prospect holes and veins of ore to interest him in backing the small operations. He listened, looked, and quietly evaluated.

Clark was experienced in mining. He had mined. He had staked other miners. He knew geology. He saw the possibilities in Butte ore, and tied up several mines. Then, to better exploit his opportunities, he went to New York and attended the Columbia School of Mines. When he returned to Butte, he was ready to build his fortune.

He built it surely and shrewdly, without, it is said, ever taking a gamble. When he died in 1925 he reputedly left a $150 million estate.

Among important early Clark holdings were the Travonia, Original, Stewart, Gambetta mines and various reduction works and smelters. He sold a large portion of his operations in 1910 to Amalgamated Copper, but retained other important Butte holdings into the twenties. In 1928, three years after his death, Anaconda bought out the remaining Clark interests. Included in the sale were: the Elm Orlu mine, Timber Butte mill, Butte street railway, Montana Hardware, *Miner* newspaper, Clark bank, Columbia Gardens, and other properties.

At one time Clark was allied with Heinze, but broke the alliance to his advantage.

He had political ambitions. He bitterly out-fought Marcus Daly to take Montana's capital to Helena instead of Anaconda. His election to the U.S. Senate in 1898 produced such a scandal that the Congress refused to seat him. Two years later he gained the seat and served six uneventful years.

He built a mansion in New York, collected a valuable art collection, but failed to break into the elite "400." The Fifth Avenue mansion was reported to be his one great extravagance. His Butte home later became a tourist attraction.

A half century of domination by one company may slowly erode the stature of other companies and men who have trod the same ground. In retrospect, possibly that has been the case with William Andrews Clark, whose many enterprises supported thousands upon thousands of Butte families over the years. Furthermore, he treated his workers well.

Whatever his motives, Clark was active in bringing Butte the eight-hour workday. He also brought the people of Butte their Columbia Gardens and Clark Park. He developed the Gregson Springs resort for them. He built the street railway system. He built and donated the Paul Clark Home to Butte's Associated Charities. He instituted group health plans for his workers. And he gave every Clark employee — thousands of them — a turkey at Christmas time.

William Andrews Clark

MARCUS DALY

Although the greatest growth of The Anaconda Company came after the death of one man in 1900, no man could be more historically synonymous with Anaconda than Marcus Daly. In like manner he doubtlessly deserves recognition for his influence upon Butte and Montana. He arrived in America a 15-year-old immigrant with fifty cents in his pocket. When he died at age 58 his fortune was vast, his peers included J. P. Morgan and William G. Rockefeller, and he was known and respected as a friend by thousands of persons ranging from muckers in the mines to Wall Street millionaires.

Marcus Daly, one of eleven children, was born December 5, 1841, near Ballyjamesduff in County Cavan, Ireland. Fifteen years later he left Ireland and worked on New York docks until he could go to San Francisco. There he met Tom Murray, another young Irishman, and they traveled to northern California mines where Daly became fascinated by mining and determined to make it his career.

Later he went to Virginia City, and the famous Comstock, which, some said, was one of the world's finest training mines for "practical" miners. Daly learned quickly and earned a reputation as an expert. The Walker brothers, Salt Lake financiers, sent him to inspect the Alice silver mine in Butte. On his recommendation they bought it. He built a mill for the mine and operated both very profitably. By the time the silver boom was over, he had discovered that Michael Hickey's Anaconda mine was rich in copper. He wanted the Walkers to buy it. They were not interested. Daly withdrew his $30,000 interest in the Alice and bought into the Anaconda.

He needed development money and turned to San Francisco financiers George Hearst, James B. Haggin, and Lloyd Tevis. He had enabled Hearst and Haggin to net $17 million on another mining deal. Daly now got the backing he needed. He secretly bought more of the hill, including the St. Lawrence, Mountain Consolidated, Green Mountain, Bell-Diamond, High Ore, Wake-Up Jim, and the Modoc. He was on his way to building the huge Anaconda empire. By 1899 Wall Street barons backed him with $75 million capitalization. Other Butte properties were acquired, importantly Heinze's, and finally Clark's a quarter century after Daly's death.

Daly's enterprises extended beyond the mines to the BA & P railway, coal in Belt, lumber in Bonner, and electricity in Great Falls. As a hobby he bred famous race horses on his Hamilton ranch. He built the Anaconda Hotel, then one of the west's finest. But towering above all else was The Richest Hill on Earth and its mines. Today the Butte operation is only part of the world-wide Anaconda Company.

Marcus Daly's impact upon Butte and Montana may be emphasized by the following extravagance of journalism from a Butte newspaper headline when he died in 1900:

THE MIGHTY OAK HAS FALLEN

The Architect Of Montana's Greatness Is Gone. Marcus Daly Is Dead; His Name And Works Held Sacred In Montana. Love That Was His Due In Life Now Made Manifest, Marcus Daly Was A Gift Of Nature. Greater Than Napoleon, A Leader Of Men, Died Amid The Monuments Of His Glory. All Montana Mourns His Death.

Marcus Daly

F. AUGUSTUS HEINZE

Frederick Augustus Heinze, 20, a student from Columbia School of Mines, arrived in Butte in 1889 and went to work for Boston and Montana Consolidated Copper and Silver Mining Company for $100 a month. He was tall, handsome, intelligent, and ambitious. Seventeen years later he left Butte reportedly worth $50 million and had waged war with two copper kings, Clark and Daly, to become the third. By age forty-five he had lost his fortune on Wall Street and was dead.

Of the three copper kings of that era, he was the upstart, the youngest, the most flamboyant, the fastest operator, the most attractive to people, and certainly the most colorful. W. A. Clark dressed with plug hat, morning coat, full and sartorially perfect beard, and dignity. He never gambled. He bought priceless art for his New York Mansion. He wore his title of Senator elegantly and proudly. Marcus Daly, the poor Irish immigrant lad who created the great Anaconda empire, symbolized the rags to riches story. While Clark collected *objects'd art* with his millions, Daly bred and raced some of the best race horses of the time, to the undying joy of betting miners. While Clark kept a discreet distance from his employees, Daly knew muckers in the mine by their first names.

Heinze created his own charisma. He drank with miners, played the piano for them, was a good boxer, and a man's man. He lived handsomely in an apartment and was one of the most eligible bachelors in the west. He was well educated, cultured, charming, adored by women, and respected by men. He also was one of the fastest men on his feet in the game of mining, high finance, and mogul-power.

In all probability the "war of the copper kings" never would have become the subject of books, articles, and newspapers if it had not been for Heinze. Those who want a play-by-play account of what happened can find books listed in most bibliographies devoted to Butte history.

Briefly, he used his $100-a-month job surveying for Boston & Montana to make his own duplicate maps of what he learned about the big hill. Within a few years he raised enough money back east (after attending mining schools in Germany with part of $50,000 he inherited) to build a smelter in Butte, cutting the smelting price in half to independent companies. He began to lease and buy mines with almost brilliant foresight. Finally he used the "apex law" to his advantage and brought great Daly and the Amalgamated Copper to their knees.

The story is excitingly punctuated with almost atrocious deals, fast-thinking attacks, employment of top legal brains in the nation, political scandal, and startling surprises.

Over all of it reigned the young man with the brilliant ability, sharp practices, astounding charisma, and the scientific education that gave him the tools to back up his defiance of older power and money. The sensible solution for full development of the great hill was single ownership because of the geological tangle of mineral veins and rights. Heinze shrewdly used that knowledge to combat Daly and get what he wanted. His was a fast, blazing, exciting ascendency in the mining camp. He took on the giants and beat them. He earned the description of "The Bold Buccaneer" (*The Copper Kings Of Montana*, Marian T. Place), and he left Butte a victor.

As one old-timer said recently, "Whatever his faults Heinze was one hell of a man!"
History can't deny it.

F. Augustus Heinze

PATRICK A. LARGEY

So much attention has been given to the three "copper kings" — Clark, Daly, and Heinze — who achieved international attention around the turn of the century that other prominent men of that era have been somewhat overshadowed. This may be especially true of Patrick A. Largey. In Harry C. Freeman's 1900 book *A Brief History of Butte, Montana* he devotes an equal full page picture to Largey along with the other three kings, and begins his sketch about Largey with this noteworthy paragraph:

> When the present shall have become crystallized into the past and a more accurate view of events shall permit, few names will stand out in such relief as will that of the late Patrick A. Largey, in connection with Butte's development. Born of modest parentage, he took into life the sterling qualities of integrity and business activity, and with these wrought out for himself a handsome heritage, besides leaving behind him throughout that life — thirty-three years of which were spent in Montana — a path of kindly deeds and ennobling examples.

Patrick A. Largey headed a wagon train across the plains in 1865 to Virginia City where he went into merchandising and placer mining. He prospered and in 1881 selected Butte for his future home. By this time he had established a telegraph line from Virginia City to Butte and through the Deer Lodge valley to Helena and Bozeman. Lee Mantle, later outstanding in Montana history, was one of his telegraphers.

Largey organized the Butte Hardware Company. He had interest in banks and in 1891 founded the State Savings Bank of Butte. He also founded the *Daily Inter-Mountain* newspaper. Old timers will remember the Largey Lumber Company. With two other men he developed the first Butte Electric Company, and was active in many other enterprises.

Among them was the Speculator mine, so named because the Largey family considered their initial investment in the mine to be highly speculative. The mine became one of the hill's best producers, and under the Largeys also was one of the most modernly operated mines in the nation.

Patrick Largey was tragically murdered January 11, 1898 at the age of fifty by a man who had lost a leg in the explosion at the Kenyon-Connell warehouse fire. Largey was part owner of the warehouse. The injured man somehow mistakenly believed that Largey was responsible for his disability.

Raymond J. MacDonald, a son-in-law, subsequently directed many interests of the Largey estate until his death in 1956. Succeeding generations of the fourth copper king, Patrick A. Largey, still are active in affairs of the state.

Patrick A. Largey

COLONEL CHARLES T. MEADER

Colonel Charles T. Meader was claimed by some to be the "true father of copper mining in the whole West." He might have been. He was a "forty-niner" in California and in 1865 he erected a copper blast furnace in Calaveros County, California.

He arrived in Butte in 1876 and bought some undeveloped claims, the East Colusa, West Colusa, and Bell. He was the first to try open-hearth treatment of ore in the camp, and he built the second smelter, the Bell.

Eventually he sold out to the larger companies. He left Butte to continue mining and was active into his eighties in the state of Washington.

Meaderville was named after him. The Berkeley pit has steadily removed that monumental suburb so that it, as with Colonel Meader, is becoming only a memory.

F. Augustus Heinze, right, and his staff in a simulated poker game. In an age of elegance, high stakes, and big business gambles, Heinze not only might hold the best hands, but occasionally he could run a magnificent bluff and make it stick.

October 20, 1903. Heinze bluffs it through. They came to assail him and left cheering him.

Heinze temporarily beat Amalgamated when his Judge Clancy declared the company an illegal combine. In retaliation Amalgamated closed its Montana operations and 20,000 workers faced a jobless winter. Heinze was blamed. The miners' union demanded that he do something about it. He said he'd explain the next day on the courthouse steps. An angry crowd of 10,000 came to listen. Heinze — once hero, now villain — used his magnificent talent for oratory and in an hour or so shifted the blame to Amalgamated and Standard Oil. The crowd cheered.

His triumph was transitory. Amalgamated persuaded the legislature to pass a "Fair Trial" bill and won the war. Heinze and Ryan, of Amalgamated, secretly met. Amalgamated principals formed a new company, Butte Coalition, and bought out Heinze on February 13, 1906 for $10½ million. Heinze left Butte.

Attorneys found their own bonanza in the mines during the war of the copper kings. Some of the finest legal talent in the nation fought long and hard in Montana courtrooms. At one time Heinze had 37 attorneys on his staff. When he finally sold out, the Butte District Court dismissed 110 suits involving claims amounting to more than $70 million.

WILLIAM CLANCY,
Butte.

Undoubtedly Judge William Clancy was elected through Heinze's support. He was big, rough, and given to tobacco chewing on the bench. Almost all his decisions favored Heinze. He was the despair of eastern lawyers and the delight of Butte miners. History suggests that he may have been one of Heinze's most valuable assets.

Models of mines, similar to this, played a large part in the apex lawsuits. As a result of the exhaustive geological study, mapping, and model-building of the great hill and area, scientific and economical development of the mines was assured for the future. One mining expert is reported to have said: "Had mining companies, as they existed in 1888, attempted to continue as individual entities, Butte would have been a ghost town in less than thirty years." Expensive as the apex trials may have been, they did enable geologists to tell the world what a treasure in copper, zinc, lead, manganese, silver, and gold the mines of Butte offered the nation.

MARCUS DALY

For years Marcus Daly stood in statue on North Main Street (left) in front of the Post Office, his back to the hill and mines as he looked south over the city. Then they moved him out to the school of mines at the west end of town.

Possibly it was a logical thing to do. Maybe a great many Butte people echoed the thoughts of an old-timer, Matty Kieley, who is quoted in *Copper Camp* as having said about the statue of Daly, a man he revered: "It is no luck will ever come of it. In life Marcus Daly never turned his arse on the mines of Butte or the miners who dug them."

Each generation has its own opinions, and who can decide the issues? Besides, the Post Office, itself, now is out on the flat. But there are those who still say that Marcus Daly looked mighty imposing in front of the old Post Office up there on North Main Street.

THREE "KINGS" AND A POLITICIAN

Cornelius Kelley, John Ryan, J. Bruce Kremer, and W. A. Clark pose in front of old Butte High School on West Park, probably in the early twenties or before. Kelley and Ryan may be considered second generation copper kings. Clark was first of all, preceding Daly and Heinze. Attorney J. Bruce Kremer was nationally active in Democratic politics. Here he and Clark appear to be wearing lapel ribbons suggesting that they might be taking a break from a political convention.

Left to right, L. O. Evans, Ben Thayer (identification not positive), Cornelius Kelley, Sam Barker, James Woodard, Carlos Ryan (son of John Ryan), Frank Kerr, James Hobbins, John D. Ryan.

This group represents some of the most prestigious and powerful men in Butte's history. Four of the men had served or would serve as Anaconda Company presidents: Thayer, Ryan, Kelley, and Hobbins. Ryan also was president of Montana Power, a position later held by Kerr. Carlos Ryan became vice-president of the power company. Woodard was president of the Metals Bank. L. O. Evans was Anaconda Company's general counsel, and Sam Barker was an ACM geologist.

ROLL CALL

These are names of Butte mines. We have tried for a complete roll call. Some may have been omitted. If so, we sincerely apologize to Butte and to those who might remember. As for correct spelling, we've selected spelling which seems to be most common. One mine may have three spellings, another may have two. Without doubt, passing years have a way of changing spellings and sometimes names. Possibly the important thing among miners was simply to identify the mine. So if over the years Mountain Consolidated became simply the Con, that was enough to know where a man worked.

Adventure	East Colusa	Maria	Rescue
Alexander	East Grayrock	Marie Louise	Rialto
Alex Scott	Edith May	Martha	Robert Emmet
Alice	Ella Clark	Mayflower	Rocker
Alisbury	Ella Ophir	Michael Devitt	Rock Island
Allie Brown	Elm Orlu	Midnite	Rockwell
Amapore	Emma	Mill Site	Rooney
Amy Silversmith	Estrella	Milwaukee	Ryan
Annie and Ida	Excelsior	Minnie Healy	St. Clair
Anselmo	Flag	Minnie Irvin	St. Lawrence
Argonaut	Fraction	Minnie Jane	Samantha
Atlantic	Free-for-all	Missoula	Saukie East
Aurora	Gabriella	Modoc	Saukie West
Ausania	Gagnon	Molly Murphy	Silver Bow
Avery	Gambetta	Moonlight	Silver Bullion
Badger	Gem	Moose	Silver Chief
Badger State	Germania	Morning Star	Silver Lick
Balaklava	Glangarry	Moulton	Silver Smith
Belk	Goldsmith	Mountain Central	Sister
Belcher	Grabella	Mountain Chief	Smoke House
Bell	Granite Mountain	Mountain Flag	Snowball
Belle of Butte	Gray Eagle	Mountain Rose	Snow Drift
Bellona	Gray Rock East	Mountain View	Sooner
Belmont	Gray Rock West	Nellie	Speculator
Berkeley	Great Republic	Nettie	Star West
Black Chief	Greenleaf	Neversweat	Steward
Black Rock	Green Mountain	Night Hawk	Sun Dog
Blue Bird	Hattie Harvey	Nipper	Sunrise
Blue Jay	Hawkeye	North Berlin	Surprise
Blue Wing	Hibernian	North Star	Tramway
Bob Ingersoll	High Ore	Ophir	Transit
Buck Placer	Jamestown	Original	Travonia
Buffalo	Jersey Blue	Orphan Boy	Tully
Burke	Jessie Wingate	Orphan Girl	Tuolumne
Burlington	Josephine	P-80	Valdemere
Champion	Kansas Chief	Pacific Slope	Vulcan
Chattanooga	Kanuck	Parnell	Wake-Up Jim
Chicago	Kelley	Parrot	Walkerville
Chinook	La Piata	Pauline	Wappelo
Clark's Colusa	Later Acquisition	Paymaster	West Colusa
Colorado	Leonard	Pennsylvania	West Grayrock
Colusa Parrot	Lexington	Piccolo	West Mayflower
Comanche	Little Minah	Pittsmont	West Steward
Cora	Liquidator	Plover	Wild Pat
Curry	Lone Tree	Poser	Yankee Boy
Cut Hand	Maggie Bell	Rainbow	Zella
Czarromah	Magna Charta	Ramsdell's Parrot	Zeus
Darling Fraction	Magnolia	Rarus	
Diamond	Manhattan	Read	
Dixon	Margaret Ann	Ready Cash	

Gray Rock

West Gray Rock

Mt. Con — 1891

Mt. View

Leonard No. 1

West Colusa

Mt. Con

Pennsylvania

Original — Mt. Con and Stewart in background

Badger

Granite Mountain

Bell-Diamond

Clark's smelter, south of the city

Berkeley

The Anaconda mine, 1926

On October 19, 1875, Michael Hickey, aided by John Gillie, staked out a quartz claim on the big hill. He remembered a Civil War editorial in which Horace Greeley wrote: "Grant will encircle Lee's forces and crush them like an Anaconda." Hickey visualized similar characteristics for his claim so he named it Anaconda.

In 1882 he sold a third interest to Marcus Daly for $15,000. Later he sold out for another $70,000. Daly got the money and a fourth interest from Hearst, Haggin, and Tevis, western mine owners.

The silver began to run out, but Daly discovered a wide vein of chalcocite which proved to be the richest mass of copper ore discovered to that time. Daly and his partners foresaw the future demand for copper with the growing use of electricity. They kept the rich strike a secret. Daly closed the Anaconda. Other mine owners panicked. Secretly Daly had trusted henchmen buy the Neversweat and St. Lawrence. Then he announced

THE KELLEY SHAFT

By the mid-'70s the underground producing mines had been reduced to the Mountain Consolidated, Steward and Leonard. Ore is trammed from them to the Kelley shaft for hoisting to the surface.

The Anaconda crew in 1900. Man circled, standing center of back row, is Tom Hosty, engineer.

his great discovery and reopened all three mines full blast. He proclaimed: "Butte is the richest hill on earth!"

His words were prophetic, but even he probably did not anticipate the eventual impact that the word Anaconda would have upon mining.

Anaconda, St. Lawrence, Pennsylvania, and Berkeley

East side of hill as it was, including the
Leonard, Tramway, High Ore, Mountain
View, Badger, Granite Mountain, and
other mines.

THE PIT

The big shovels took their first bites out of Anaconda Hill in 1955. By the 1970s the Berkeley Pit was considerably more than a mile square and 1200 feet deep from the western side. It was producing about 300,000 tons of material daily, using 133 huge trucks, ranging in capacity from 100 to 200 tons. Nine of 15 power shovels could take 15-yard bites. A fertilizer, ammonia nitrate, mixed with diesel fuel substituted for dynamite in 40-foot blasting holes, with 200 holes making the minimum blast. Dozers, graders, rubber-tired loaders, fire truck, ambulance, and wrecker for emergencies were in use.

The big pit, the largest copper mining pit in the country, also had crept to about four blocks from the center of town.

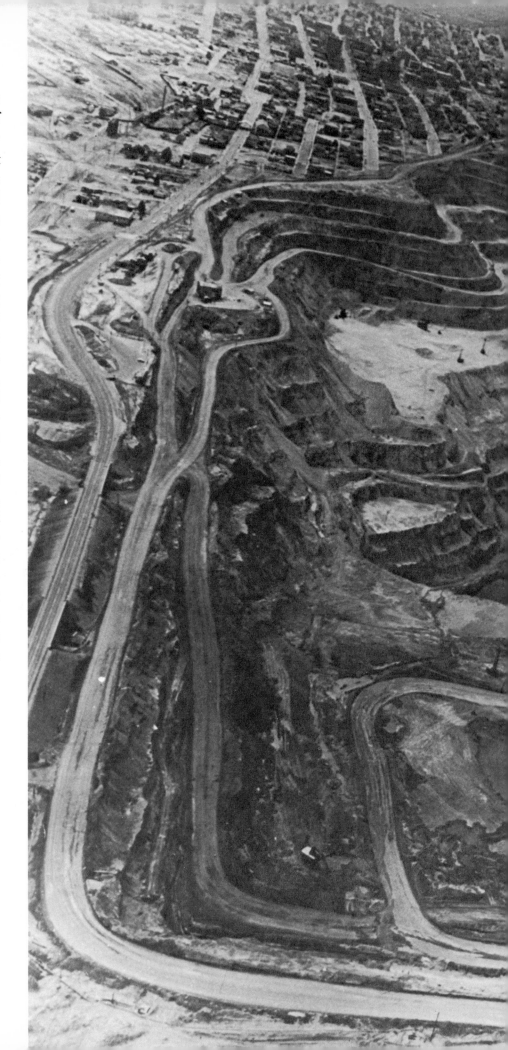

But others didn't find pay dirt

THE CITY

City views belong to different generations, each particularly conscious of its own vistas, and each in later years remembering how it was. Cities have been built, changed, and destroyed by man. They have been destroyed by nature. They have been destroyed simply by time.

So Butte has changed. Yesterday's Butte is not today's, nor will today's be tomorrow's. Because this book is essentially a trip back through nostalgia, we are including very few contemporary pictures of Butte. Possibly, in a way, we are depicting a city of another time.

Nevertheless, every generation, including today's youngest, will recognize places and buildings and streets and houses that still defy weather, time, and even *man*.

It takes a strong city of character to do that!

THE PICTURE . . .

Almost out of a dream was this artist's concept of Butte as he visualized it in 1904.

Close study of his picture reveals houses and buildings in astonishing detail. Streets are defined. Missoula Gulch bisects the West Side. The great hill with its works is dominant. A familiar west wind bends plumes of smoke toward the mountain range.

The people must be there because the picture somehow teems with activity. A piquant realism attests to the city's very existence.

Yet the artist seemed somehow to do his picture as he saw the city in his mind's eye, from a desire to create his own city from the real city, and to make Butte the city it was, but wasn't, and perhaps could never be — except in the hidden thoughts, emotions, and dreams of all who ever loved it for what it was, and could never quite be.

Now it is almost gone, but there is Butte as an artist pictured it in 1904. Perhaps, too, there it is as some still remember it. Passing years have a way of extravagantly enhancing memories.

IN THE 1870S AND 1880S

Early pictures may be indefinite as to precise identification or time, but the mountains are a constant reminder that this is Butte.

Main Street from Quartz, 1877

ANACONDA HILL AND LOWER BUTTE, MONTANA.   C.R. SAVAGE, SALTLAKE.

1881-1882

Butte Montana — 1882

A historical Butte scene

Caption by C. Owen Smithers, Sr.

"This view of Butte from back of the Methodist church on North Montana Street shows many early landmarks. In the immediate foreground is the fancy steeple of the first Courthouse. Clocks, set in all four sides, could be seen from most of the city. Back of the Courthouse, left center, is the first home of Marcus Daly, the copper king. Later it was a home for the Central High School girls. Also in the foreground is the first Mountain View Methodist Church steeple. In the distance, left, is the Parrott smelter. Far out, right is the Clark Bell smelter.

"This is an exceptionally good picture inasmuch as smoke from the smelters usually was so thick that only on days when the west wind blew strong could distant places be seen. The scene probably was photographed about 1895."

82

In the 1880s

Park and Main in the 1890s

For years the hub of the city was "Park and Main" which served as a streetcar terminal, meeting place, and focal point from which east side, west side, north side, and south side originates.

In Butte's heyday thousands lined up for the latest movies at the Rialto theater on the southeast corner. On the southwest corner was the Metals Bank in the city's most impressive building. On the northeast corner was the Owsley building, later renamed, and subsequently and more recently destroyed by fire. The Lizzy Block on the northwest corner never achieved great distinction, except as a familiar landmark.

Owen Smithers remarks about the above picture, taken before the Metals Bank building was erected: "In the distance can be seen the dense smoke from smelters. A picture of this kind could not be taken unless a strong wind is blowing. The smoke from the smelters on the outskirts of Butte was so bad that one could not see across the street in the middle of the day."

Today the smelters are gone. So is a substantial part of Butte's old "Park and Main."

Butte hill in the mid-1880s, west side

Butte began to come of age in the 1880s. In 1882 it had a population of 4,000. Then Marcus Daly and W. A. Clark began to develop the mines. Butte boomed. Hundreds of miners rushed into camp. By 1885 the population was about 22,000. Passenger trains were arriving on the new Oregon Short Line from Ogden. Butte out-did copper production in the famous Lake Superior region.

The early mines on the big hill usually had covered shafts. Groupings of miners' homes clustered near the important mines. In the downtown business district a steady increase in the number of substantial brick buildings attested to a settling-in permanency that, in some cases, would possibly exceed a century of occupancy.

East Broadway, about 1895. The big building to the left is identified as the "Big Ship" where a constant stream of miners lived.

Butte in the 1890s was distinctly a camp for true hardrock miners. Automated machinery still was to bring its full impact to the drifts and stopes. Drilling was hard work with hammer and hand-held steel. A muck-stick was the proper tool for loading an ore car and mules lived most of their lives underground pulling the cars through the workings they seemed to know even better than the men.

Coming off shift and down the hill to a boarding house such as the Big Ship, or to a house or flat and family usually meant facing the discomfort of heavy smelter smoke.

Artists' sketch collection

Four pages are here assigned to Butte sketches collected by C. Owen Smithers, Sr. who found merit in all graphic references to places, people, and history.

The West Park scene to the right obviously is somewhat more contemporary than those below and on the following pages.

East Side of Main Street, from Broadway to Park

North Side of Granite Street, West from Main

North Side of Park Street, from Main East

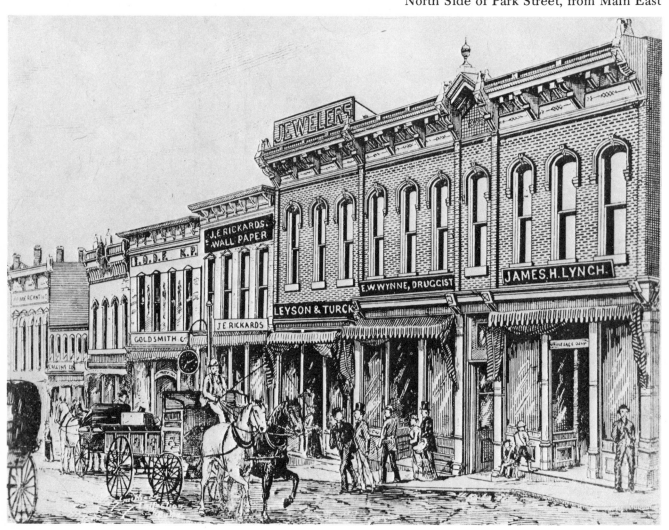

West Side of Main Street, from Quartz to Granite

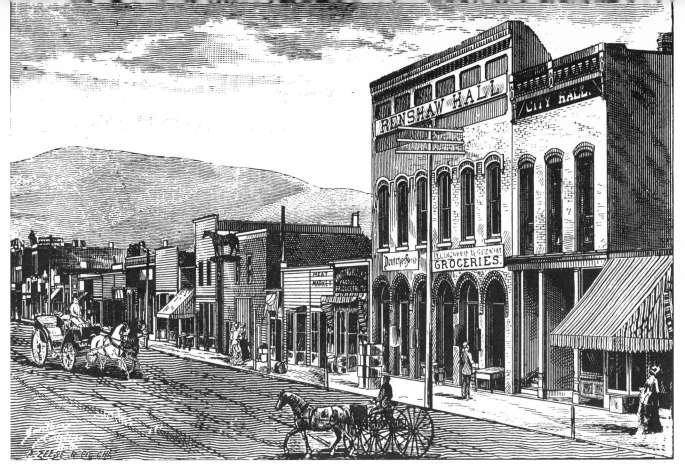

South Side Park Street, from Academy East

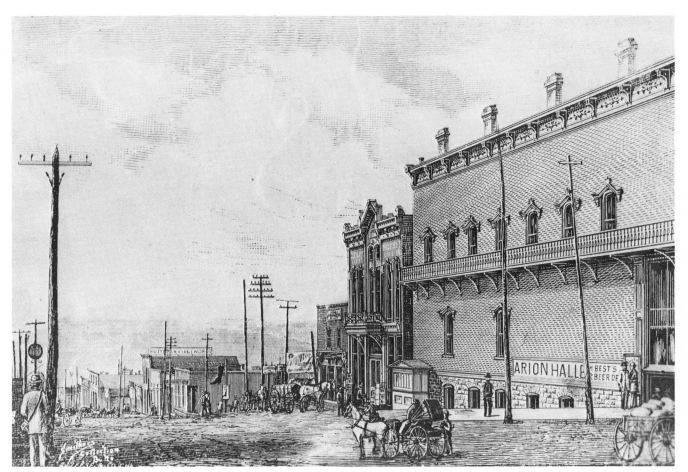

West Side of Main Street, from Park South

1-ALICE HOIST BLDG   7-MEAT MARKET
2-ALICE MILL   8-STORE
3-ALICE OFFICE BLDG 9-ALICE BLACKSMITH
4-ALICE STABLE   10-ALICE CARPENTER
5-ED REIMEL RESIDENCE 11-WELL
6-SCHOOL HOUSE   12-WASTE DUMP
           13-MOULTON MINE
           14-BELLE OF BUTTE MINE
           15-CLARKS FRACTION MINE

WALKERVILLE, MONT,
    OCT 1878.

Smithers
Collection
Butte

Suburb — 1878

Centerville was close to the mines before the day of the big pit

Dublin Gulch

Meaderville — a place to remember

For years, and especially beginning with Prohibition, mention of Butte anywhere in the world usually evoked a comment about Meaderville, which once nestled at the east base of the big hill, crowded against mine fences, until the Berkeley pit began to engulf the 70-year-old suburb.

This was the famous Italian settlement where food, wines, nightclubs, modest homes, backyard gardens, and friendly, gracious old-country families were legendary to thousands upon thousands who visited or lived in Butte. Possibly the finest recorded tribute to Meaderville was written by J. H. Ostberg for his *Sketches of Old Butte.*

For years Meaderville was a settlement with two images. At night its main street, pictured here, was crowded with visitors who stayed far into the small hours enjoying what was called, during Prohibition, "The Little Monte Carlo." Nightclubs offered delicious food, good drink, gambling, and excellent dance bands.

A second image lay behind its main street among the homes where, at one time, there resided about 2,000 persons, most of whom lived in the old, traditional, Italian family culture. Ostberg described it this way: "A family-life that is shared by all, where none are slighted or ignored, where the happiness of one becomes the happiness of all, and the sorrow of one is sincerely felt by everyone, comes pretty close to achieving the thing called civilization."

Meaderville certainly was a place to remember.

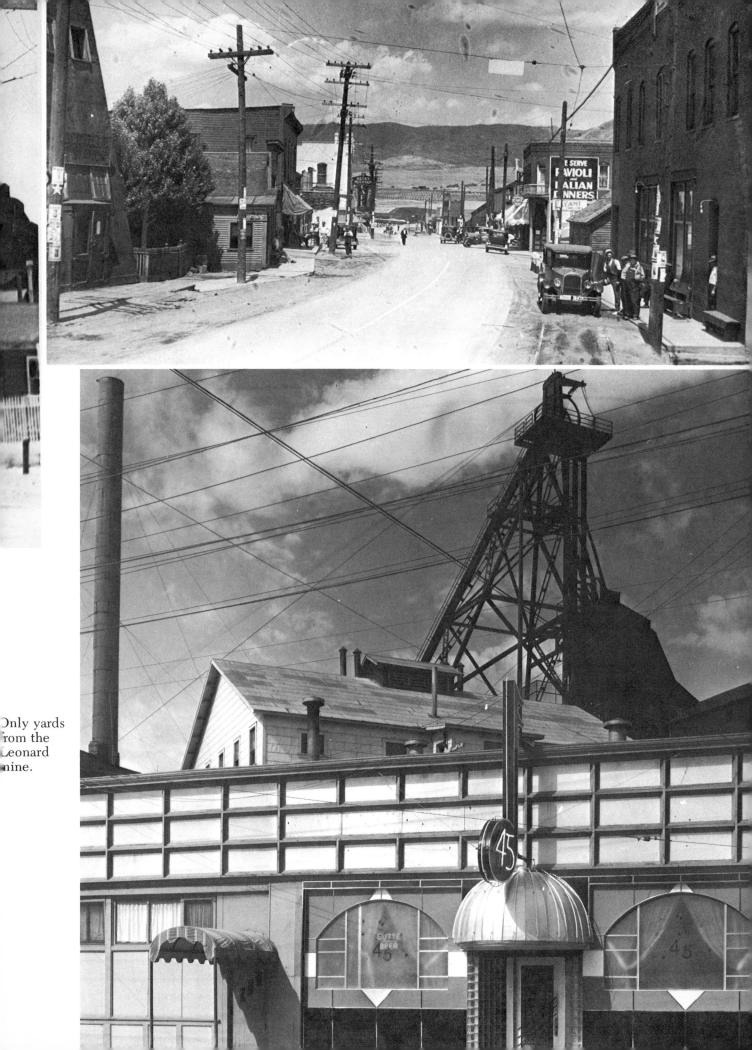

Only yards
from the
Leonard
mine.

The old Cabbage Patch occupied several blocks a short distance from "the district."

From a distance the housing project that displaced the Cabbage Patch resembles a new patch on a battered face.

The unglamorous end of a once glamorous
way where a thousand girls entertained
thousands of men . . . and they called it
Venus Alley.

The Mecury Street entrance to
Venus Alley had major parlor
houses on each side.

The Copper King, at Galena and
Wyoming, was never a brothel,
but many girls of the line lived
there.

Mountain man

Turn counter-clockwise to see the profile of the "old man of the mountains" east of the city.

Butte winters can be very cold. Days in the 20s, 30s, 40s, and even 50s below zero are to be expected. Nights are colder.

BUTTE AT NIGHT

Untold thousands have exclaimed over the view of Butte at night. Historian Stewart Holbrook once said that at twilight from the mountain pass where the Northern Pacific approaches the city: ". . . Butte presents a dramatic picture, twinkling with astonishing brilliance in the high thin air." He then added that when viewed in the light of noon, ". . . the big camp is as ugly as sin and just as fascinating."

However true Stewart Holbrook's words might be about the noontime view, few are unimpressed by the distant night view of Butte.

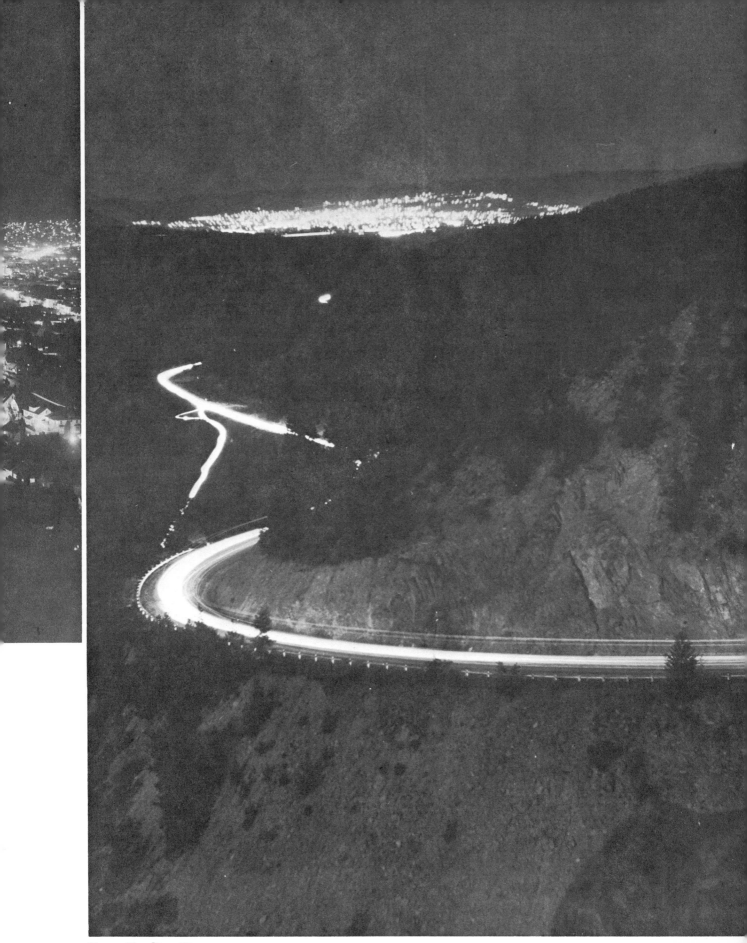

From Harding Way

Each generation has its own memories of how it was . . .

STREETS OF BUTTE

In old Butte most streets are on a hill, especially those running north and south, and not all those running east and west escape a hill or two somewhere along the way.

At various times these hilly streets have inspired uses other then commonplace traffic. Kids on roller skates have found the sidewalks to be so fast that many have used a length of discarded broomstick braced against a leg for a brake. Bicycles have become speed demons going down, but very tough pumping going up. Some streets once had streetcar rails. Slippery rails from ice and snow could be a problem on a steep hill, and on Halloween rails might be mysteriously and somewhat dangerously greased.

Possibly one of the best unorthodox uses for the longest and steepest streets has come with winter. Some streets have been blocked off for streets and sleds — especially in the old days. With a well-seasoned Flexible Flyer, a short running start, a belly-flop on the sled, and a kid could ride with gathering speed for block after block, mittened hands tight on the guide bar, sub-zero cold and snow stinging into cheeks and eyes. Then it was a long climb back, but the breathless ride down was worth it. Or, if luck was good, a grocery delivery wagon might be going up the hill, and every sled had a rope to get a ride behind wheels.

Most spectacular have been the bobsleds, usually long, heavy planks for maybe up to eight or more passengers, with steering sled runners mounted up front, and stationary runners at the back. A clanging bell cleared the way. They could start up in Walkerville or Centerville and clang an alarming, spectacular, exhilirating, hell-bent-for-election descent down the long hill to a selected up-hill or level stretch and a safe stop.

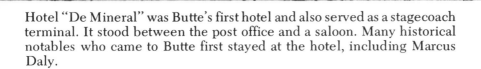

Hotel "De Mineral" was Butte's first hotel and also served as a stagecoach terminal. It stood between the post office and a saloon. Many historical notables who came to Butte first stayed at the hotel, including Marcus Daly.

Scenes and Places, Happenings and Faces

Butte never was a place to mature calmly or beautifully. Some cities may be referred to in the feminine gender, but it seems that Butte always has been masculine. And as the face of Butte weathered, it had no beauty of the lovely woman to lose, but rather it has been the rugged aging face of a man that reflects a hard life and great strength of character. Butte always has been a man's town, but it never would have been much of anything without its women.

East Park Street 1887

Car barn of Butte Cable Street Car line in Centerville. Cable cars coasted down and were pulled up by cable.

Cable car line

West Park Street in front of Masonic Temple

Uptown corner, 1899

Front Street mud

Waiting across from Hennessy's store and the company's famous "sixth floor"

Looking west on Park showing old, familiar names, including the Peoples theater on the left and the Owsley block on the right.

More familiar old stores and names looking east on Park from Montana

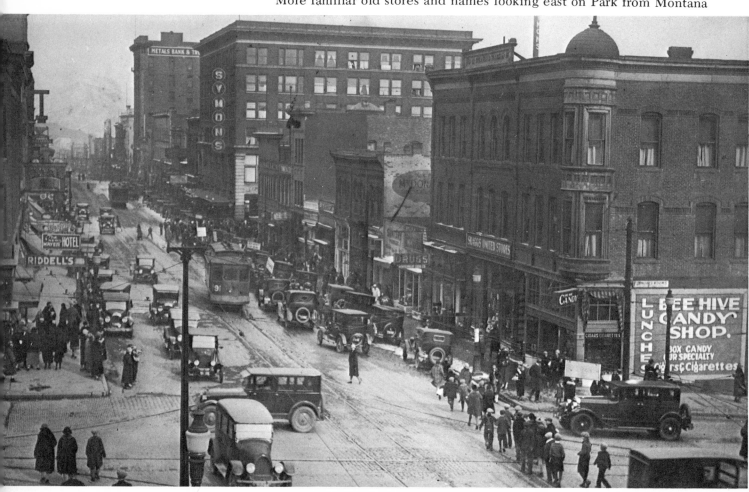

Main at Park before the Rialto was built

North Main from Galena and more names, including a Lutey store — one of the first grocery chains in the nation

PARK STREET

When Butte ran high in population and activity Park Street probably was the most travelled downtown street, and certainly West Park was closer to being the city's "white way" than some others.

Looking West, 1928

West Park toward Park at Main with the Finlen Hotel looming in the background on East Broadway.

Looking east from West Park at Montana.

Park at Main in the early 1930s

124

West Park in 1941

West Broadway

Broadway, Looking east

North Main Street from Park at Main before they moved Marcus Daly's statue
from in front of the old post office.
    In 1973 fire destroyed most of the
landmark buildings, first block, right, including the old Owsley building that for
years housed the Butte Business College.

American Theater fire.

Clifford's and Butte Hotel fire on East Broadway

The comparatively recent devastating fire on West Park wiped out several of the city's landmarks.

130

FIRE

Over the years Butte fires have taken heavy toll above ground as well as underground. In more recent times, fires have plagued the city center to literally erase much of "old Butte." Fire destroyed a large part of Columbia Gardens before power shovels could move in.

The Elks celebrated years ago, Broadway at Main

Christmas scene, East Broadway.

Underground workings shifted the surface and curved streetcar tracks on
South Montana.

Murray Hospital

St. James Hospital

The funeral

Butte's first motorized hearse

The old courthouse

Library before the fire and remodeling

THEATERS

Butte always has been a great show town, especially during earlier days. Butte loved show people and show people loved Butte. Those who appeared in the mining camp would make a most respectable list of greats. Once Butte supported its own opera company under dapper John Maguire in his Grand Opera House. Repertory theater reached its zenith under direction of "Uncle Dick" Sutton.

Of somewhat less propriety were music halls. The most notorious was the Comique that stood on the Metals Bank Building site. For five years, 1893-98, the Comique never closed its doors. On the main floor were tables, drinks, girls, and entertainment. On the second floor balcony were private boxes with lockable doors and pass-through panels for drinks. Ingeniously painted screens allowed occupants to view the excitement below without being seen. Wealthier young blades of the city came in a back way to occupy the boxes with their ladies of the evening. The vaudeville and burlesque stage shows featured top performers. Other music halls flourished, but none as sophisticated as the Comique.

When motion pictures took over, the city was plentifully supplied with theaters. The Rialto, managed by Billy Sullivan, held top honors, but others also played to full houses on weekends. Vaudeville was popular and attracted the top circuits. New York roadshows regularly played in the city, usually at the Broadway.

Most theaters are gone. The city has the usual complement of residential television antennas.

"Uncle Dick" Sutton, center with cane, poses in front of the Grand

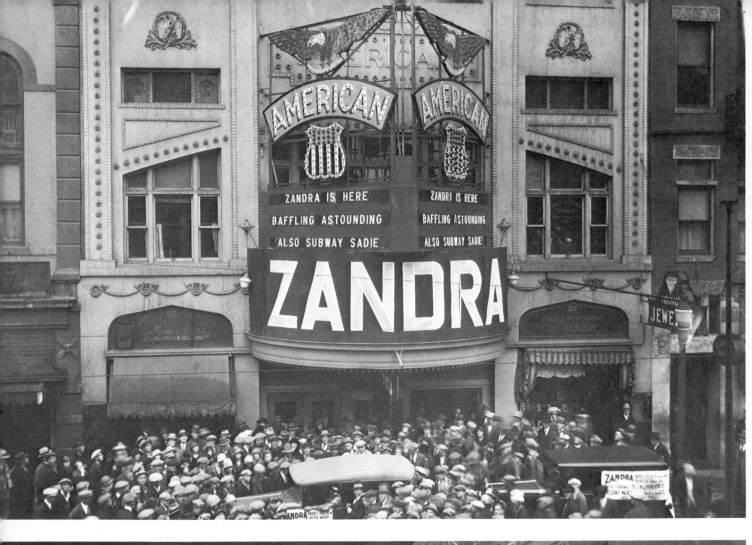

OTHER PLACES

Out-of-town recreation always has been popular in the city. Night life has offered a series of highway stops including the old Bungalow, and the Five Mile and Nine Mile houses. Family picnics and more sedate outings and recreation attracted many to Lake Avoca where boating was in vogue. Gregson Hot Springs has been the site of natatorium and outdoor swimming, Northwest swimming meets, countless picnics, and one rip-roaring riot, August 20, 1912, when thousands of Butte miners and Anaconda smeltermen clashed after a tug-of-war event. A vast supply of empty beer bottles became effective weapons. Women and children fled to the hills. The B. A. & P. dispatched a special baggage car to take the more seriously wounded to hospitals.

The Five Mile

Gregson Hot Springs

The time of the big bands reflected the great popularity of "ballroom dancing." The Winter Garden for years was crowded on dance nights, and some of the nation's top bands played there. A "battle of music" between a visiting recording band and the outstanding Winter Garden band would more than pack the house.

"The Reservoir" picnic grounds has entertained organization
picnics ranging from school classes to firm employees such as
the annual Symons picnic pictured here.

The Butte Y.M.C.A., completed after World War I, made a quick impression upon the city in the early 1920s under Sam Parker, one of the nation's top money-raisers and Y executives. Mel Clevett, who later became a Y.M.C.A. international executive, directed an athletic program that soon encompassed thousands of persons. Almost immediately the Y became more a community center than the customary Y, and women were invited to membership and use of the gym and pool. Thousands of kids enjoyed free Y privileges. Organizations, including Girl Scouts, Boy Scouts, Camp Fire Girls, and other civic groups made full and constant use of the meeting rooms.

Athletic directors Clevett, Oscar Bjorgum, Oscar Dahlberg, Ben Hardin, B. J. Gunderson, Marian Fitzpatrick, and Dorothy Cloke were instrumental in developing the citywide programs for school children in conjunction with Larry (Lala) Manion of the Knights of Columbus and other participating groups. During this period, the top three floors were almost exclusively the residence for unmarried mining, business, and professional men. The doors to the building never closed.

On the street floor Ed Freitag managed the pool and billiard room and eight bowling alleys. He established some of the best bowling leagues in the nation.

With passing years the Y changed, possibly reflecting changing lifestyles in the city. In the spacious second floor lobby imposing stained woodwork was painted and the lobby "modernized," partitioned, and converted to other uses. Some old-timers say that they notice other changes — that although the Y still has a significant place in the community, it's not on quite the same high activity level of its first three decades. On the other hand, the community, with its needs, also has changed.

When the copper kings reigned, the Silver Bow Club became one of the nation's noteworthy men's clubs. Eastern guests of importance were astounded by the decor, cuisine, wines, bar, game rooms, living quarters, and impeccable service.

With changing times the elite club ceased to exist as a club. In the Great Depression the building housed some government relief agencies. The metamorphosis was completed when it eventually was occupied by the miners' union.

The Knights of Columbus has contributed much to the city's recreation and character growth. Working with the Y.M.C.A. the organization and its facilities assumed a heavy load to provide its share of athletic and recreational activities for Butte children with the inception of the city-wide program in the 1920s.

The old Country Club in the Lake Avoca area where one of the state's first golf courses was laid out at the foot of the Rockies.

Smithers
Butte

Early picture —
in the old days simply called
Montana School of Mines.

Old Butte High School

More recent picture —
now elaborately called Montana College of Mineral Science and Technology!

Fat Jack Jones, a Civil War veteran and Butte's first and last hack driver, was internationally known. Presidents, kings, actresses, champions, and many famous persons of his era rode in his hack. For more than 50 years Butte knew him and he knew Butte. Finally he retired to a Soldiers' Home in California and died in 1920 at the Los Angeles home of W. A. Clark, Jr. Some historical sketches about Butte grant him almost as much attention as they give the copper kings.

The Fez Club, purchased by Butte Shriners in the 1940s, was built by one of W. A. Clark's sons in 1898. He imported the mansion from Europe, each piece carefully numbered for assemblage in Butte. Its many fireplaces, woods, ironwork, furnishings, and artifacts have gained high value as antiques. Freeman's 1900 history of Butte pictured it as C. W. Clark's home.

The Marcus Daly country home, Hamilton, Montana

COLUMBIA GARDENS

In 1899 W. A. Clark purchased land at the foot of the Rockies and established Columbia Gardens for Butte people. In a smelter-ridden city they needed it.

No one who ever lived in Butte since then to about 1974 — three quarters of a century — needs to be told about Columbia Gardens. Probably every kid who grew up in Butte during most of those years knew about Children's Day when transportation out there was free, and every Thursday during the summer was great.

When W. A. Clark sold out in the twenties, he also sold the Gardens. "The Company" kept up the tradition for years. In 1973 things changed. Despite efforts of citizens to "save the Gardens" the corporation killed an "image" to get the copper.

On the other hand. W. A. Clark, Marcus Daly, Con Kelly, and men like that probably saw things differently.*

*For those who want more, we recommend *Memories of Columbia Gardens* compiled by Frank Quinn.

154

Children's Day

157

163

THE PEOPLE

Waiting for a parade at Park and Main, 1926

THE PEOPLE

The people of a roaring, uninhibited, working mining camp are said to be like no other people. If Butte has been the greatest in history — and there are those who say that — let there be full and appreciative recognition of the generations who helped to make it so.

They knew the streets of Butte where they walked tall and proud, confident, arrogant, humble, gentle, mean, beautiful, or ugly; but above all as men and women who set their own values upon work and life and death.

They were mining barons and muckers. They were gentlewomen and they were girls from the line. They were preachers and gamblers, merchants and politicians, doctors, lawyers, brawlers, and cops. They met eye to eye, worked shoulder to shoulder, lived and died — each in his or her own way — and they made Butte the world's great mining camp.

They attended Chautauqua at Columbia Gardens. They bet on horses, dogs, cards, dice, baseball, football, and fights. They set a rip-roaring pace through Prohibition, and they flocked by thousands to a revivalist's tent. They suffered strikes, depressions, wars, and mine disasters with courage. They rejoiced, celebrated, paraded, and rioted with fervor. They poured money into programs for their kids, supported their churches, patronized their saloons, fought for their unions, died for their mines, cursed their city — and fervently loved it.

Now in a new era, older generations might not completely recognize themselves in the misting mirror of memories, nor be fully understood by their more youthful progeny, but let there be no mistake about the people of Butte: they have been always magnificent.

They rejoiced, celebrated, paraded, rioted — watched flagpole sitters — with fervor. Looking toward Park and Main one gala day, probably in the '20s.

Nostalgia — Butte boys and caps

Nostalgia — Butte men and hats (William Gibbs McAdoo, speaker, early '20s.)

Nostaliga — Butte girls and garlands at Columbia Gardens

PRESIDENTS, POLITICIANS, AND OTHERS

It would appear that Montana's electorate would never be numerous enough to greatly affect national politics, but what the state may have lacked in quantity, it has more than made up in the effectiveness of its elected representation. Montana Senators and Representatives in the Congress have more than made themselves felt in the national and international scene. Montana politicians have made history on many occasions, and Montana men and women have been appointed to high office in government.

Always politically conscious and demonstrative, Butte, itself, has entertained an imposing number of presidents, candidates, behind-the-scenes politicians, and other celebrities. Butte also has contributed significantly to their ranks, and through them to the nation's history.

"My friends . . ." Franklin Delano
Roosevelt, 32nd President, was im-
mensely popular in Butte.

Upper left:
Theodore Roosevelt,
26th President.

Butte people turned out to hear presidents and candidates. An estimated 10,000 persons listened to FDR. A portion of the crowd is shown above.

Lower left:
Harry S. Truman,
33rd President.

William Howard Taft, 27th President (center)

John F. Kennedy, 35th President, with Mrs. Kennedy and high school students

Warren Gamaliel Harding, 29th President, at a Butte mine

Dwight David Eisenhower, 34th President, stopped to talk in Butte

William Jennings Bryan — ran for President, 1900

Charles E. Hughes — ran for President, 1916

Alfred E. Smith
ran for President, 1928.

Wendell L. Wilkie
ran for President, 1940
(Mayor Charles Hauswirth, right).

Senator Burton K. Wheeler (6th from right)

Senator Mike Mansfield (center)

Robert Kennedy ran
for President, 1968.

Thomas E. Dewey ran for President, 1948

Lindbergh came to Montana, 1928

Will Rogers

Lindbergh

Rogers visits Butte, 1926

Senator Wheeler and Dr. Thomson of the School of Mines

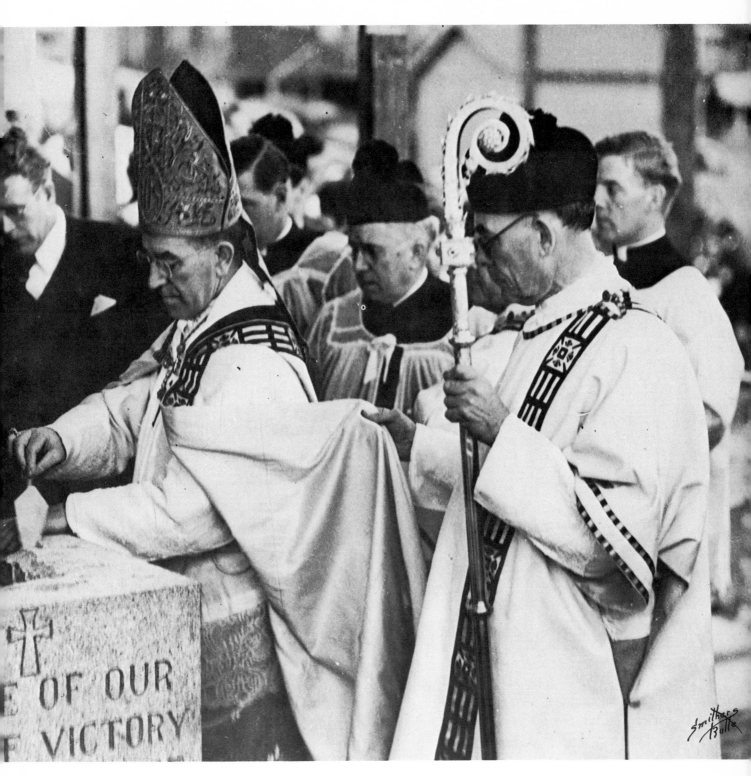

A cornerstone is laid

"JERE THE WISE" — BUTTE'S FAMOUS CHIEF OF POLICE

One of many, many stories goes that some early-in-the-century gang-sters had registered into the Butte Hotel. There was a knock on the door. They opened it and faced a six-foot, two-hundred-pound, well-dressed man with eyes like steel and a face hard as granite. He said, "My name's Murphy. There's a train leaving here at three o'clock. See that you're on it."

They knew — as the rest of the nation's underworld did — that he was Jere Murphy, Butte's famous Chief of Police. His exploits were fabled, his ability uncanny, his love of Butte unmatched, and his integrity unquestioned. They were on the three o'clock train.

Jeremiah Murphy, born in Ireland, spent most of his life as a distinguished law enforcement officer, and died September 19, 1935 as the result of a fractured skull when he fell in a scuffle to disarm a crazed man who was terrorizing the Montana Power Company offices a few doors from the police station.

TONG WARS IN CHINATOWN
Tong wars were not uncommon in Butte's old, well-populated Chinatown. Jere Murphy finally put a stop to them after a particularly brutal murder of a Chinese poultryman. Weapons and a relic or two from opium dens are displayed above.

Nickel Anne

Uncle Dick Sutton

Straight Back Dan and his newsletter

Lemmons was often seen balancing a loaded tray headed for customers in "the district."

Shoestring Annie, a famous mining city character

Two ladies of the Cabbage Patch

191

Butte loves a parade . . .

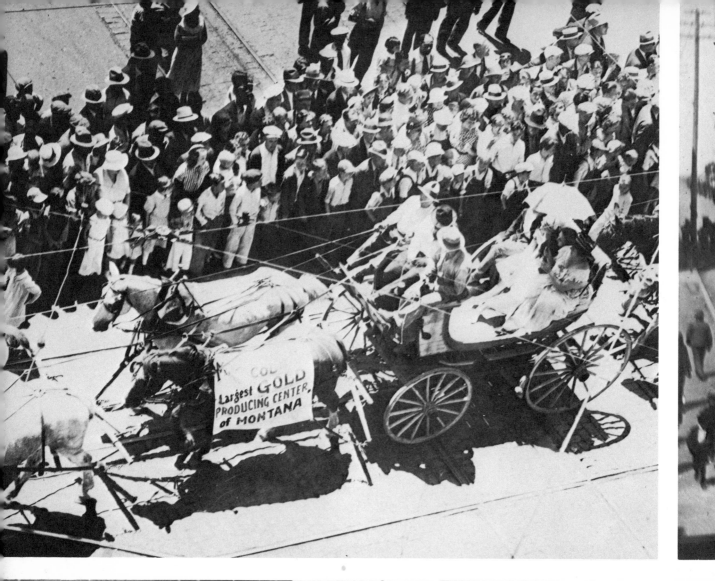

Armistice Day, 1922

Sometimes the aftermath could out-do the parade. The "A.P.A." riot on July 4, 1894.

The Famous Butte Mines Band - 1937
Sam Treloar - Director

Sam Treloar's Butte Mines Band for years was one of the city's institutions. It played for parades, funerals, concerts, and important occasions. Occasionally it went on tour and was enthusiastically received wherever it appeared. The musicians were recruited from the mines and other Butte enterprises. Treloar was a gifted musician and a superb bandleader.

First inspired in the late twenties by Connie Belle Hoover, Camp Fire Girls director, the annual Doll Parade became an annual and delightful event for Butte's small girls and thousands of people who have lined the parade routes to applaud.

Groups

rd Annual Meeting
Royal Order of Jesters
celebrating
Chas Millers birthday party
Sept 4-5-1932

CALVES MAY COME.
AND COWS MAY GO.
BUT THE BULL
GOES ON FOREVER.

Smithers
Butte

Tom Larson-Chouteau   7- Geo.Gossman-Dillon
M.Burlingame G.Falls   8- Tom Scott-Dillon
James Spear G Falls   9- Lew Adams-Dillon
-Angus McLeod-Butte   10- Sam Shiner-Butte   14- Geo. Lovell-Dillon
Frank Conley-Deerlodge   11- John Hertzler "   15- Geo W. Dart- Dillon
O.L. Dillenbeck-Anaconda   12- S.C. Small -Deerlodge   16- Alf.Whitworth-Deerlodge
                           13- Sam White - Butte   17- Frank Mulch -Butte
                                                         18- Dr.R.W.Getty - Galen

19- Smith McKnight-   Dell
20- Dr. Chevigney -Butte
22- Casey Pierce   "
23- Don Anson-Wisdom
24- Charlie Miller   "
25- Dr.H.D. Kistler - Butte
26- Alex Christie   "
27- Dr.Harry Bolton-Warm Spr.

201

"LINE UP AND SMILE!"

Untold thousands of groups want to be pictured for one reason or another, and countless photographers face the task of photographing them.

Group pictures probably contribute little to graphic drama or excitement. Usually they commemorate an event, a happening, or simply a photographic document that a group exists. For the moment, at least, the group is important enough to have a picture taken. Each person in the group is important to that person, usually to a family, and possibly to many others. Thus the group picture may have a measure of historical and personal significance.

We think that is sufficient reason to include the following pictures of Butte groups — especially for those who may recognize some groups and faces.

Some pictures have little if any identification of persons, time, or reason for being. Some do. To tell the complete who, what, where, when, and why of each picture would be an almost impossible task. We dip far back into Butte history and its people. We ask you to forgive us our shortcomings and mistakes in captions. We have done our best.

W. A. Clark and friends

Indians frequently camped out on the flat

Ladies of distinction

Veterans of the Civil War took part in the
dedication of the Civil War Cannon at the
Gardens in 1904

Butte's Civil War Veterans 1930

Smithers
Butte

Civil War Vet9
1930

Co. G. FIRST REGIMENT MONT. VOLUNTEERS, U.S. BUTTE/98.

y-3-1929

Wm L. Grayson
Commander-in-Chief
of
nish-American War Vets
sits his Comrades in Butte
ne group are standing around the capstan of the Battleship Maine

Smithers

atte's First Quota of the First Draft — Left Butte Sept 5/1917

1 — Lawrence Keegan
2 — Leonard W. Warner
3 — Clifford F. Harris
4 — Harold W. Grory
5 — Chester Richards
6 — G. L. Zilkey
7 — Ellsworth DeSnell
8 — Fred Root (Back)
9 — Tommie Dowd
10 — Lloyd Heilman
11 — Julian M. Savage
12 — Wm Mondlock
13 — Arthur Forrest
14 — Richard B. Born
15 — Geo. R. Carle
16 — Wm Daniels
17 — H. F. Weyerstal
18 — Anthony L. Schmidt
19 — August G. Lyssow
20 — Clarence N. Smith
21 — William Grey
22 — Ham R. Taylor
23 — Earl K. Miller
24 — Ettore Giovanetti
25 — Alexander Taft
26 — Wm O. McCormick
27 — Clement Hyde
28 — Emil K. Ozanne
29 — Edward J. Edwin
30 — Mayor Maloney
of Butte

Copies of this photograph may be obtained from Smithers Butte

E. Cook 37 Robt M. Ryburn
Steese 38 John R. Evans
Nerugg 39 G. C. Smith Alternate
Woll 41 Percy Fordyce
42 Leonard Estlick
Collins 43 Van Leningham
R. Gross

BUTTE AND THE WARS

The people of most American communities always have faced wars with courage, sacrifice, and honor. If one community may differ slightly from another in performance of its people during the wars, those differences may stem from the people themselves, and their lifestyles. Possibly the people of some communities are better fitted through civilian life to meet the trauma of war. This may be especially true of Butte simply because for years men of fighting age came from the mines where physical danger and death always has been a daily occupational hazard.

Thus when the men of Butte have gone to war, many of these men have been somewhat immunized to danger and death, as have their wives and families. By the same token, many of these men of the mines have learned a self-competency in the face of danger. This often makes for extraordinary stamina and courage.

Should this also contribute to heroism and unusual bravery and performance in battle — or strong support and courage on the community homefront — it may simply be part of Butte and its people. The war record of Butte people has been outstanding and enviable.

The last veteran

Men of Distinction

Byron H. Sewell Floyd Ike
Cook Davis Closser Congdon

William John Ed
Gimmell Charles Congdon

D.J. William
Charles Wilson

Bulle Harri

Struthers
Collection

214

Butte businessmen in earlier years. (Identification not positive.)
Front row, left to right: L. S. Shonninger, unidentified manager of
Hennessy's, Ruben Hobbs, Jacob Cohen, unidentified. Back row,
Sig Schilling, John Sullivan, John L. Hannifin, Ben Meyer, Dan
Colemen, Sam Shiner.

Early directors of the Y.M.C.A.: Tom Davis, Dr. Donovan, Dr. Witherspoon, "Ad" Carroll

The 4 Jacks was the camp's first "exclusive" club, and possibly the forerunner of the Silver Bow Club. Pictured here about 1887, it stood on North Main. Andrew Davis was president.

Four prominent fishing cronies: Ben Calkins, who operated an office supply store, Thornton, who built the Thornton Hotel, Charlie Henderson, once sheriff and Malcolm Gillis, who had a coal yard. They were partial to the Madison river. President Hoover fished with them.

At one time the city had a penchant for electing blind men to justice of peace and judicial office. They became widely known for fairness, possibly because they accepted the credo that "justice is blind" for their own dedication to office.

217

Butte Police
Old Timers
1933

Walter
Morrison
F H White
James
Doolin
W. Jack
Ingraham
Joseph
Powell
Joe
Williams
John
Barray
Pete
Harvey

Butte police, 1933

Jere Murphy, center, in his
earliest days on the force.

Highway patrol on
North Montana.

**BUTTE'S FIRST PAID FIRE DE-
PARTMENT**
First row: George P. Porter, Tom Coberly,
Wesley and Charles Barnaman. Second
row: W. E. Ross McDonald, Fred Dug-
dale, Edgar Daylon. Third row: William
O'Malley, William Scolly Orr, John Sloan,
Frank E. Young, George Fifer.

Sheriff's Posse, while legally
only an echo from the old days,
has been active in Butte.

READY TO ROLL

Early horse-drawn firewagons and buggies made spectacular runs through the city. The city's firefighters have a long history of ability and bravery. Fighting fire when the thermometer dips into the forties and even the fifties below zero calls for unusual, dedicated men. The department has kept abreast of modern techniques. Older buildings in the city have created staggering fire fighting problems.

THE GREAT EXPLOSION

Historically, one fire always has been of note. On a cold winter night, January 5, 1895, the Kenyon-Connell Company warehouse on the south side near the upper yards of the Northern Pacific caught fire. It was stocked with mining supplies, including 350 boxes of dynamite.

In three shattering explosions the warehouse disintegrated and killed all but three of Butte's firemen and scores of bystanders. Total casualties were unofficially set at sixty. Only one fire horse survived and later became a city pet until put in pasture until his death of old age.

THE FOURTH ESTATE

At times Butte newspapers headlined their own history in a fashion that would delegate most contemporary media to mediocrity. The copper kings and corporations owned newspapers. The unions controlled their own. Libelous accusations frequently scorched front pages, although libel suits apparently were not the order of the day. Power politics, the war of the copper kings, the violent conflicts between labor and management all contributed to some of the most colorful journalism the nation has witnessed.

Newsmen came to Butte, wrote, left, and became famous elsewhere. Some stayed and helped employers try to control the city and occasionally the state. As the Anaconda Company gradually dominated much of Montana's economy, it also sought to dominate the newspapers. When the Clark interests sold out in 1928, The Anaconda Company owned the morning *Anaconda Standard*, published in Anaconda but distributed mostly in Butte, and the *Daily Post*. Clark's *Miner*, a morning newspaper, was Butte's oldest and possibly most widely read publication. For years it had usually voiced a corporate editorial trend along with the *Standard* and demanded that the company "retire from politics, abandon its foolish campaign to choke publicity through its costly chain of newspapers and attend strictly to its own business" in its July 26, 1928 edition.

The company promptly bought the *Miner* and combined it with the *Anaconda Standard* into the *Montana Standard* so that Butte had two major company newspapers, the *Standard* and *Post*. W. A. Clark, Jr. promptly began publication of a new morning newspaper in Butte, the *Montana Free Press*, one of a chain of three publications he formed. He employed most of the old *Miner* staff. The following year the *Free Press* suspended publication, leaving Butte again with the company *Standard* and *Post*.

On June 1, 1959 the Anaconda Company got out of the newspaper business when it sold its chain of newspapers to Lee Newspapers, a midwest chain based in Mason City, Iowa. Included in the sale were the *Anaconda Standard*, *Billings Gazette*, Butte's *Montana Standard* and *Daily Post*, *Livingston Enterprise*, the *Sentinel* and *Daily Missoulian* of Missoula, and the *Helena Independent Record*.

At one time Butte supported four daily newspapers: the *Miner*, *Standard*, *Inter-Mountain*, and *Evening News*. Labor published the *Bulletin* and later the *Montana*

Joe Markham   Butte Tom Wigal
Bill Clark   Jack Zygmond   John Colcater
Bert Gaskill   Jean Jordan   J. D. Murphy
Harold Seipp   Nick T.

ess Club — 6 May 1956

c Ward T.L. Walt Glen Frank T.C Dick Wm J.D
y Fanning Greenfield Nelson Moon Graff Greenfield Daniels Quigley Holmes
Colin Ed Owen DeVan Wayne Mike Richard Al
Raff Coyle Grinde Shumway Farley O'Connor "Shag" Miller Gaskill
Chester Robert Alex J.H. Law Absent
H. Steele Cavagnaro Warden Dickey Risken Al Gusdorf
nk Quinn Clayton Maxwell Louis Poole Ross Hagen Tom Maddox Ed Hanmer

Labor News. In the 1930s the outspoken *Eye Opener* made its mark. Butte's newspaper days of flamboyant headlines and bitter competition are long gone, but when Butte, itself, frequently was making international news, the news was more than adequately covered by some of the best journalistic talent in the country.

225

Years ago newsmen often were imported from the best eastern newspapers to man staffs of Butte newspapers owned by the copper kings. Unfortunately no specific identification can be made of the journalists here pictured. Probably none of them would sanction running a picture without identifications no matter how impressive the picture. Consequently we apologize to them — wherever they are . . .

Meanwhile, the admittedly incomplete and somewhat vague identifications are as follows: Front row, left to right: Spike Haines, Louis Thayer, Wally Walsworth, Dolph Heilbronner, "Dapper" Dan Conway, Joe Gilbert, Bradford St. Charles, Jim Kelly, Harry Seltzer. Second row, Griswell, Whitacher, Foster, Elsworth, M. M. Miller, Dick Kilroy. Third and fourth rows (incomplete), John McIntosh, Bill Cheeley, Pete Nelson, Colonel Searles, John Cole.

Senator
Harry
Gallaway
Larry Frank   Tom Greensfield   H.B.Dunshee   P. Gay
Dobell Roach   Jack   Dave   Stivers   John Corette,Sr
                    Walter Hamill W.A.ClarkJr charles J.K.Heslet   J.H.Dursten Eddie Hanmer
Charley "Brownie"   Harold   H.R.   Alex Divine Chas Shearer   Dudley Richards   L.Q.Evans
Copenhaver Bill Floto Brown   Crary chas   Whorton   De la   F.L.Furgerson Louie Thayer
ns   Jim Cummings Cohan   John Kieth   Alex   Wally
        Gov Sam Stewart   Senator Ed McLaughlin London   Harry   Johnson Walsworth Ned
                          W.A.Clark   James   H.W.Wallisen Goldfarb   Jim Phillips "Spike"
                                            Berry   John McIntosh   Hair

CLARK ENTERTAINED NEWSMEN

Senator W. A. Clark invited prominent Butte and other Montana men to his annual breakfast at the Silver Bow Club. Owen Smithers probably said it best in his caption: "To be invited to this breakfast set one aside as someone of distinction in the newspaper game in Montana. In this picture is Governor Sam Stewart standing next to Clark. He also invited men who opposed him in many activities; the lieutenants of the Anaconda Company, and the opposing newspapers in Butte, and especially all the heads of his many activities."

227

The *Butte Miner*, owned by W. A. Clark, was Butte's first newspaper. At one time it shared a much smaller building with the *Daily Inter-Mountain*. Its tenancy in this West Broadway building lasted until the newspaper was sold and moved in with the *Post* on South Main.

E. G. Leipheimer, pictured here between Father Leonard, left, and *Montana Standard* reporter "Bing" Rooney, right, with journalism students of Central High School, went to work for the *Post* in 1914, and became managing editor of the *Montana Standard* when the papers consolidated.

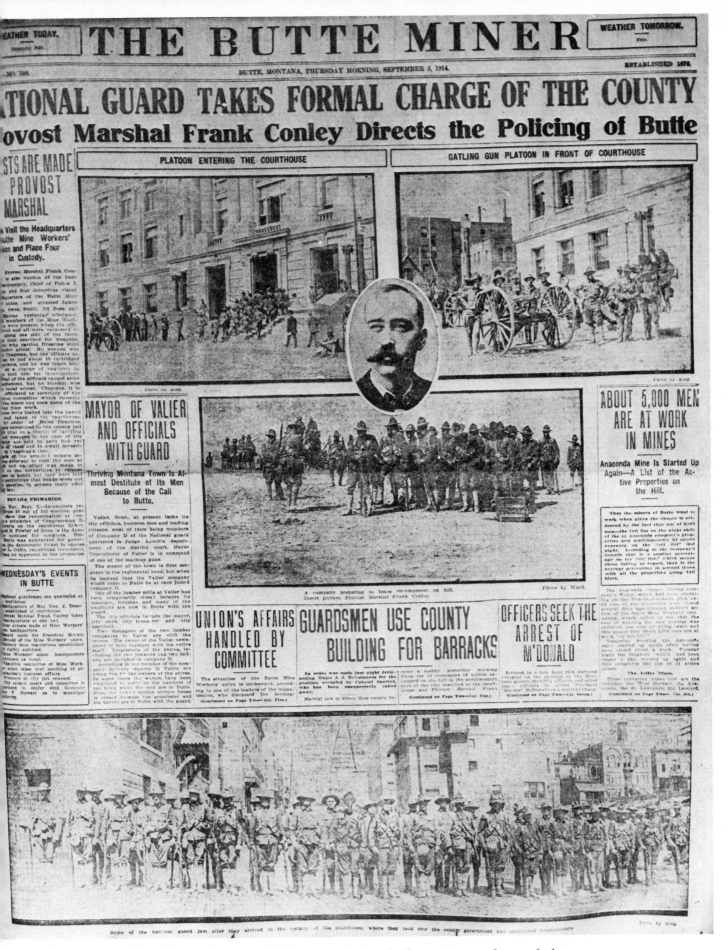

Front page of the *Miner* when martial law was declared after union warfare ended in violence.

Obviously posed with tongue in cheek, but the background speaks for itself and the faro table was very much for real. The gent, posing in derby, was C. Owen Smithers, himself.

If the rate of hardrock production was high in the old days when thousands of miners were at work, the rate of hard liquor consumption was proportionately high when the miners were off shift. As many as 700 saloons have been reported in early times and it is said that a man coming off shift could drop his bucket on many a bar, slap down two-bits and get a shot of whiskey and free beer chaser, possibly as a gesture of congratulations that he had survived one more shift underground.

The friendly neighborhood saloon. Notice two men with buckets for beer.

In 1919 the 18th Amendment brought
Prohibition to the nation and almost
brought a major upheaval to Butte, traditionally a hard-drinking mining camp.
Law enforcement officers poured liquor
into gutters while crowds looked on.

Bootleg stills operated in or near the city. This one was raided. Butte resisted Prohibition with vigor and about 500 speakeasies.

In 1933 the nation decided that Prohibition was a failure. The amendment was repealed. Breweries and distillers began to operate again. Finally old-fashioned draught beer arrived in Butte.

First Keg beer recieved by Walkers April 8-1933 Butte Mont

Smithers Butte

Happy day in Butte!

The only place in the
United States
that served
Draught Beer
over the bar
April 8-1933
Walkers
24 hour service for 35 yrs
Butte Montana

Smithers
Butte

Girls from the old Catholic Central High School, 1928 — across the street from the Mt. View Methodist Church on North Montana.

236

A city for all people, all faiths, all orders, all ages.

Lodge members in front of Butte's Masonic Temple — in upper left corner part of the Knights of Columbus building.

Honor Day to Rotarians
of 15 years service or more
Dinner at Hotel Finlen
Dec 12-1933 - Honor guest
Fred Bennion-District Governor

Smithers Butte

DAUGHTERS OF ST. GEORGE PICNIC
GREGSON SPRINGS
JULY 30 1925

More Butte Kids . . .

Tony and some of his thousands of Admirers in Butte

William S. Hart, movie star (*never* "Bill" Hart), was a hero in
early Westerns, and could roll his cigarettes with *one* hand —
shoot deadly with *both*. Lucky were the Butte kids who *met*
him!

Arbor Day in 1904 was typical of many such days faithfully
observed by school children in the area. Hundreds of Arbor
Day trees planted over the years stand today as silent memori-
als to the passing generations.

Some of these pioneers may have been in the Arbor Day picture of 1904.

Serbians in Butte

St. Patrick's, one of Butte's oldest schools

JOSHER'S CLUB

The Josher's Club is said to have originated in Al Green's saloon on North Main one December night when Billy Gimmel, a gambler, saw a small boy shivering by a radiator. He proposed that the men at the bar begin a movement to give the poor people in town a good Christmas dinner. They did and for a quarter of a century the Josher's Club continued to raise thousands of dollars to give Christmas dinners to needy families whose identities were kept secret.

Tim Kearney   Billy Gimmel   Archie Coutts

Butte's famous Rocky Mountain Cafe in Meaderville was the background for many banquets where Teddy Traparish served world renowned food.

Montana State Hairdressers Assn.
935 Convention — Butte Montana
Banquet
at
Rocky Mountain Cafe
in Meaderville

—Butte High School Girls Glee Club—
1939-40

Smithers
Butte

BUTTE MINES BAND

52nd Anniversary
Banquet at the
Rockey Mountain
Cafe

Sept 11th

1939

Percy   George   Antone Geo.   Bill
Percy Vogwill Wrightson Leshovar Stevens Treloar   Bill Castle
                                              Owen Lloyd
Joe Rich   Edwin James, Jr.
Frank Holly   John Sacomano

Herb Davis   Clarence Johnson
Maurice Moore   Henry Kubish
                    Howard Rich   Howard Kitto
              Tommie Gribble   Ted James, Sr.
         George H. Wilcox   Jack Robbins
       Trevor Thomas   Dennis Sullivan
    Harry Temby   Al Kreitenger
Tom Jenkins   Ben Ivey
Sam Treloar - Bandmaster

One of the city's first Boy Scout troops

Tonsorial perfection was offered in John Jarvis's barber shop in the Hotel Finlen basement. Note the individual shaving mugs arrayed on the end wall.

In the early 1900s a successful new well out on the flat was important enough to bring this group. The site is thought to be near Amherst at Farragut and the gentleman in the derby may be Mayor John MacGinniss.

249

Depression days — Butte boys at a CCC camp

A later generation — Camp Fire girls at Camp L. O. Evans, Georgetown Lake

Butte's annual civil war: Butte High vs. Central. The warrior in the center, without helmet, is identified as Danny Hanley, a Central man, who later played for Knute Rockne at Notre Dame.

Before radio or television covered big sporting events, newspaper sports editors bellowed blow-by-blow or play-by-play descriptions through megaphones from newsroom windows. Many will remember Sports Editor Jean Jordan's, "Flash!" The breathless pause, and then: "He's down! One — two — " Or during a world series: "Flash! It's a home run!" And the crowds cheered.

SPORTS IN BUTTE

Few cities in the world have loved sports as intensely, exuberantly, or extravagantly as Butte, especially in its early days. A mining camp is essentially a man's place. Life could be rough and tough. Violence was common. Death stalked underground. Work calluses were matched by hard knuckles ready to fight and men could bet on any contest.

They bet on horse races, dog races, Gaelic football, Cornish wrestling, boxing, animal fights, cock fights, football, baseball, basketball, hockey, handball, and bowling. The Scots curled. The Italians played boccie. And on Lovers' Roost one summer while other kids played baseball on The Cinders, an Irish kid named Kilroy taught Cousin Jack kids how to play cricket!

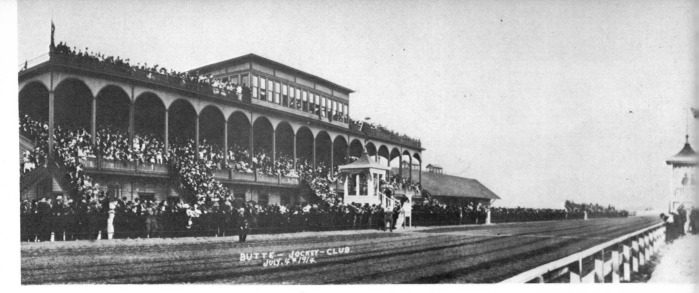

Marcus Daly owned some of the finest race horses in the world. Their winning performances are legendary. By 1897 Butte and Anaconda race meets were the best in the west. In that year $2,225,000 passed through the pool boxes at the two tracks.

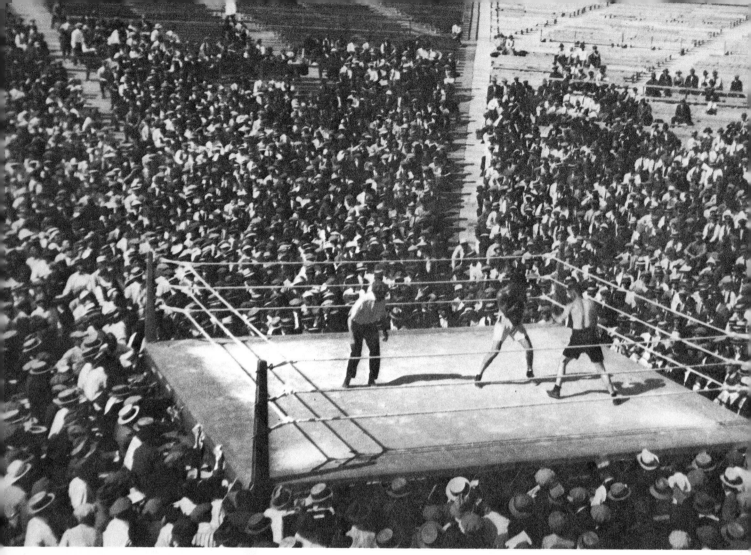

Dempsey-Gibbons world heavyweight championship fight, Shelby, Montana, July 4, 1923.

THE FIGHTERS

Butte always was a fight town for the pros. Champions came and fought. Local fighters graduated into national fame. It was not uncommon to have two fight cards and once two world championships on the same day. John L. Sullivan, Kid McCoy, James J. Jeffries, and Stanley Ketchell appeared here. The names can't be listed without probably missing many. Nevertheless, we list these fighters, out-of-town and local, who knew Butte. If you're a fight fan, you'll recognize many. Or you might consult the record books. They also tell a story.

Jimmy Britt, Kid Broad, Dave Cusick, Jack Dempsey, Leo Flynn, Joe Gans, Mike Gibbons, Jack Grimes, Ike Hayes, Aurelia Herrera, James J. Jeffries, Jack Johnson, Spider Kelly, Stanley Ketchell, Buddy King, Mose and Sil LaFontise, Dixie LaHood, Harry Maguire, Jerry McCarthy, Kid McCoy, Eddie and William McGoorty, Joe Mudro, Jack Munroe, Battling Nelson, Jack O'Keefe, Thor Olsen, Jimmy Quinn, "Boy" Robinson, Joe Simonich, Maurice Thompson, Kid Tracey, Joe Wolcott, and a stable of Sullivans, including John L., Mike, Dan, Montana Jack, and Twin. Other names identify the accompanying pictures.

256

Stanley Ketchell
as he appeared in
Butte Montana —
about a year before
he began his fight
career
Photo from an
old Tintype taken
about 1900.
The shorter of the
two is "Little Murphy"
a flyweight also of
Butte ———
Copy by Smithers
Butte

Stanley Ketchell

CHAMPIONS

Champions and ex-champions have known Butte well, have appeared as guests, and in the ring. The four fighters pictured here came to see a heavyweight fight, had dinner together, and this may be the only picture showing them as a group. Fight fans will recognize, left to right, Tony Galento, Barney Ross, Gene Fullmer, Henry Armstrong, and Joe Louis.

McDonald – McCoy
1887   At the old arena on the 'Flat

Smithers
Butte

Jay   Troy   Lou   Kenny   Joe   Sandy   Jack   Jack   Charley
Evans   Evans   Fontana   Davis   Simonich   Sadler   Archer   Campbell   Johnson
Referee   Ex-Fighter   Ex-Fighter   Fighter   Referee   World Champ   Commissioners   Sadler
Ex-Fighter   Mgr.
                                        Sonny O'Day   Al Livingston
                                        Com Ex-Fighter   Sec'ty Com

Butte Mont. Group taken right after Legion sponsored K. Davis-S. Sadler fight - 5th Apr '55

BUTTE'S LOVE AFFAIR WITH BASEBALL

Probably baseball players have never played on faster fields than some of Butte's where no well-groomed turf eases the speed of grounders. A hard-hit ball may come to earth on little more than bedrock granite and wild bounces develop great fielders.

The history of the camp's teams and leagues is long and colorful. Professional, semiprofessional, and amateur baseball has attracted a ready supply of fans — usually as partisan and vociferous as any that ever packed a stand. For years the big leagues were dotted with players who had served their time on Butte fields, and the various Butte leagues had their share of old timers coming down as well as young players going up.

Probably most exciting was the era when the big companies supported the Butte Mines League and fielded some of the best teams the city ever saw. Many players came from out of state, but native Butte players more than held their own. Feelings ran high, and once a professional umpire sought refuge in the Y.M.C.A. from an angry mob calling for his blood after a disputed call in a crucial game.

The Young Muckers
Butte Montana
1929

Ball crowd at Clark's Park, 1924

1913 — The champion Berkeley Mine team

1927 - Montana Power baseball team

1927 - Clark baseball team

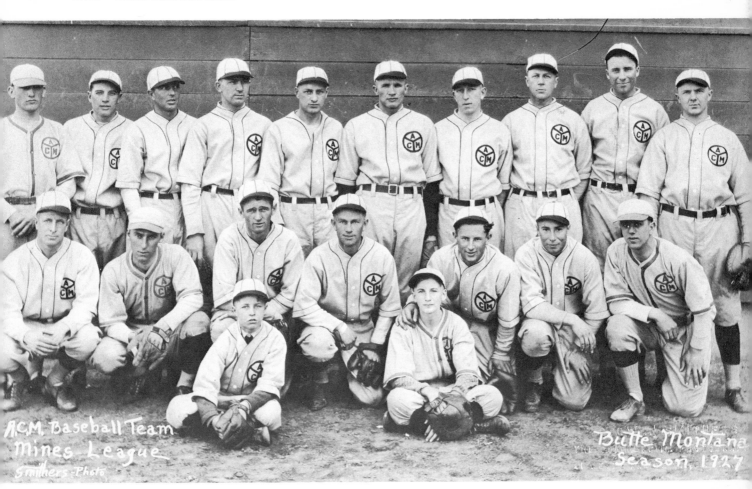

1927 - ACM baseball team

FOOTBALL

Contact sports and Butte fans seem to be a natural combination. Although the sports are tough and rough for football players — American, Gaelic style, or soccer — it often wasn't too comfortable for spectators with weather well below zero and mountain winds threatening frostbite.

Much play was at Columbia Gardens, then shifted to Clark Park, and then to Butte High stadium in 1938. School and independent league teams have always offered Butte hard and good football. Fans have a record of being highly partisan, and more than once the action in the stands has been as hard-hitting as on the field.

Professional football began for Butte in 1893. This team was labeled "Butte Professionals, 1895 — World Champions." With no national leagues in that time, no network television, no 100,000-plus spectators at a game, no records kept, and no super-bowl — who's to say that they weren't?

As early as professional football came to Butte, high school football was in the news. This was the Butte High team, in 1893-1894.

Northwest Champions
Butte High 1907

Duke Schroeder — Edgar Wild — Bud Leonard — Art Saner — Chris Nissler — Johnnie Evans — Bob Hopkins — Dick Roach — "Tub" Wilcox — Billie Dezell — Joe Phillips — Fred Brooks — Rusty McIntire — Irwin Evans — Paul Williams

Stars from this team continued to play professional football for years

By 1926 the Butte High squad looked like this

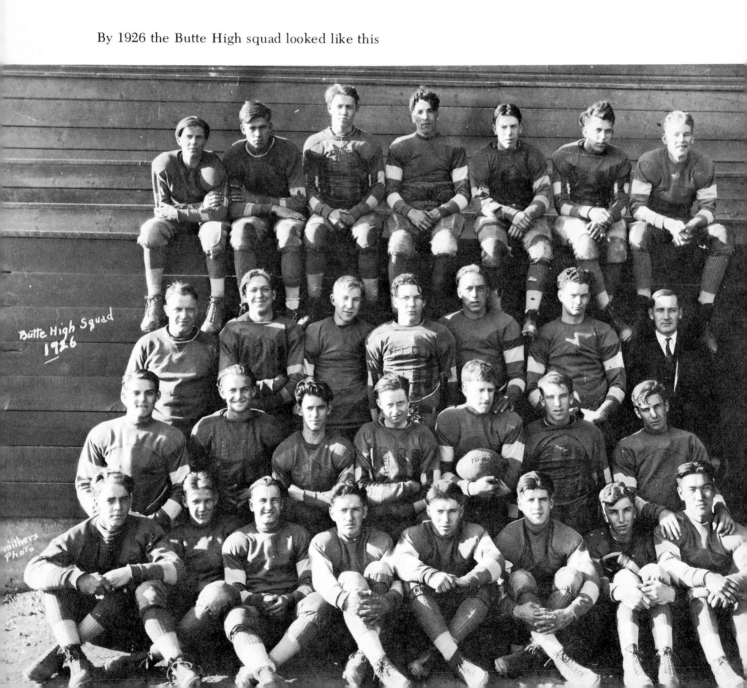

Butte High Squad
1926

milhers
Photo

Smithers Butte

28 Apr 1956 — Members 1922 Butte High Football Team at "Swede" Dahlberg Dinner
John Cox   Jim Bertoglio   Frank Grady   Carl Tysel   Jim Lyman
Mickey McMahon Frank Jaquelle Harry Dahlberg Ed Molthen   Charlie Davis
                                   Coach

Many honors went to "Swede" Dahlberg
who coached great teams and was feted at
an appreciative dinner in 1956.

DUBLIN GULCH FOOTBALL SQUAD-BUTTE 1923

Frank Conley	"Tip" Boynan	Pat Dennehy	"Gravel" Leary	Walter Marron	"Shimmy" Crnich	Shon O'Neill	"Bishop" Leary	"Butler" Driscoll
	Hugh Quinn	"Coke" Dennehy	Art Goodman	Puss Peihale	"Yonska" Tomich	Frank McKelvey	Harry Tevlin	
		Neil "Flat" Murphy	Walt Smith	Jerry Lowney Mascot	Bill Thrasher	Danny Cohan		

Centerville Champs Independent League 1914

Bob Darragh   Slug Sullivan   Otis Lee   Roy Bray   Beadle Kelly   Baldy Holland   Doddie Sullivan   J.J. O'Leary   Curly Sullivan   Babe Thomas   Kack O'Connell   Tommy Kurz   John McNulty   Jim Combs   Bidda Rogers

Walter Scott, in officiating shirt, accepts one more honor at an early Butte-Central high school game. He played All-American football for Harvard, then came to Montana and eventually the School of Mines. For years he guided state high school teams in their sports and leagues, picked all-state teams, and became an authority on Montana history.

Members of this early soccer football team in Butte are unidentified. The game was brought over from the old country and was popular. Butte teams are said to have met all challengers in or out of the state.

Hockey

Butte Copper Leafs   Butte Champions 1955-56

Coach
"Tubie" Bill   Bob   Tom   George   Larry   Dan
Johnson McManus   Knievel McManus   Bertrand   Nelson   McLeod

Bob   Jack   Darrell George   Jack   Jerry   Bob
Neary   Neary   Parke   Bronsen   Bonny   McGivern   Burton

Eddie   Barry   Floyd   Ernie Jack   Bob   Tom   Lou n This Line
McManus   Whalen Halvorsen Smith Grace McLoughlin Hitchcock Smith Absent

League bowling got off to a great start in the '20s at the Y.M.C.A. under Eddie Freitag. Mudro's was only one of many outstanding teams that would hold their own in any league today.

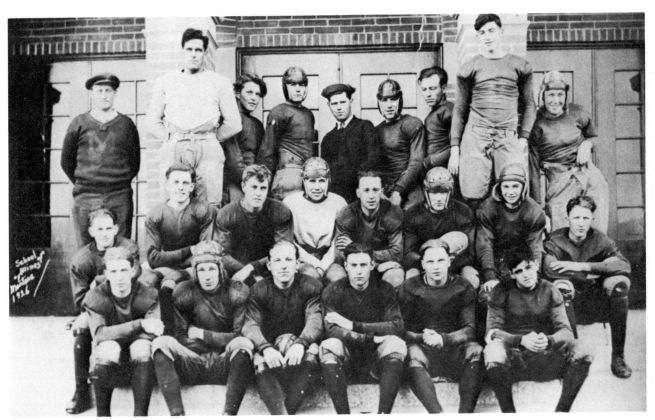

The man fourth from the left, back row, is identified as "Pokey" Powell, an end on this School of Mines team. Later he became the Lone Ranger in the movies, was a marine in World War II, and died heroically in an attack landing on a South Pacific beach.

BUZZIES 1950 FOOTBALL SQUAD - BUTTE MONT.

Back row: M. Hampton, A. Cutler, J. Tomich, J. Kello, R. Edwards, G. Huddleston, Bo Stanich, J. Masonovich, W. Clifford, G. Lilly, F. Annala, W. McKecknie, W. Boston. *Standing:* M. O'Leary, C. Stears. Mgr. R. T. Tutty, Monk Semansky, Coach R. Sparks, Line Coach L. Connors, Mgr. C. Judd, Mgr. M. Mulcahy, W. James, Joe Matkovic. *Sitting:* Miller, C. Shontell, H. Watts, J. Bennett, M. Casick, L. Mann, J. Kovacich, J. Harris. *Front row:* J. Berryman, unidentified, J. Scown, A. Chavez, J. Gasper, J. Morrisey, G. Twardus, M. Weathers, F. Donovan, missing . . . Vern Dirie.

Basketball

State Basket Ball Tournament
Butte Civic Center
19-20-21 Mar 1953
Saturday nite crowd

Always a crowd . . .

The "Bobcats" of the Butte Boulevard · 1927

Gene "Bull" "Beano" "Rusty" Bob   John   Carl   Blanchard John
Hamry Flynn Ferkovich Holland Holly   Merzlak Deshler   Lubick.   Ambrosetti
                                                                      Coach

Bob   "Babe"   Nick   Frank   Jimmy   Timmy   Dick
Harley   Ferlan   Gayaski Gardner Keane   McClough O'Malley

Their uniform was one lone pair of cleated football shoes and one worn old football helmet.
This bunch of kids only lost 1 game in 3 years

And always the kids

A THOUSAND FOOTBALL PLAYERS

The kids of Butte never have lacked for energy, enterprise, enthusiasm, or entertainment. Over many years much of the entertainment was self-generated and sometimes notably undisciplined. The rugged and colorful characteristics that have marked the camp's adult population have been also abundantly evident among the juvenile residents. As a writer observed in *Copper Camp,* "It means something to have been a kid in Butte."

By 1926 the kids had such a fierce inter-gang rivalry that the situation was becoming serious. The neighborhood gangs that once had shown some admirable qualities in leadership, self-organization, loyalties, and other attributes were changing. In writing his thesis for a master's degree years later, Lawrence J. (Lala) Manion, a key worker in Butte youth organizations, summed up the situation:

A fierce inter-gang rivalry resulted in survival of the fittest attitude. In inter-gang athletic competition the losing team nearly always criticized the inexperienced officials' decisions and a free-for-all brawl resulted. It was almost impossible for a member of one gang to go into the territory of another. Lack of equipment for athletic contests resulted in many injuries and also fostered petty thievery on the part of the gang to secure some of this much needed equipment. Gang leaders led their followers in rock fights against rival gangs in which property damage and bodily injury often resulted.

Juvenile law enforcement officers made many arrests and the number of offenders sent to the state industrial school was on the increase. This approach accomplished little, and when conditions continued to grow worse, Mayor Horgan and Alderman A. Wilkinson, Chairman of the Parks and Playground Committee, called a meeting of civic clubs and interested youth organizations to organize a program to remedy a very bad condition. At this meeting Mr. Ben Harding and Oscar Dahlberg, physical directors of the Y.M.C.A. were appointed with the writer (Manion), who was physical director at the Knights of Columbus, to outline a recreational plan and submit it at a later meeting.

The effort was highly successful. Butte established a program for kids that gained national attention. During the 1926-27 school year 18 public and 9 parochial grade schools approached full participation in organized activities for boys and girls. Employed high school boys also had a program. Scores of adults volunteered as referees, judges, and coaches. The Y.M.C.A. and Knights of Columbus furnished gymnasiums, swimming pools, and paid supervisors.

Thousands of kids participated. A report showing sponsors and events in the first year reveals what the program offered: Kiwanis Club, football, softball, baseball for boys, fieldball, softball for girls; Rotary Club, basketball, horseshoes, kite flying, marbles for boys, roller skating, jackstones for girls; Anaconda Copper Mining Company, ice skating for boys and girls;

Exchange Club, track for boys and girls; Y.M.C.A. and K. C., swimming for boys and girls; Rotana Club, basketball, O'Leary, hop scotch for girls.

Expanded activities in Camp Fire Girls, directed by Connie Belle Hoover, Girl Scouts, directed by Julia Greiner, and Boy Scouts directed by Paul Campbell rounded out one of the most active youth programs in the United States.

The *Montana Standard* evaluated the program in an October 3, 1930 editorial, "A Thousand Football Players":

> Among educators and moralists it long has been the conclusion that play-time for youth is the danger-time. Butte's fine development of organized play under a method of operation, which has been gradually perfected through four years of continuous effort at all seasons of the year, is making play-time distinctly profit-time for boys and girls. All in all, it is another demonstration of Butte's active interest in its children. The excellent leadership furnished through the efforts of the "Y" and the Knights of Columbus is transforming this interest into an active and far-reaching influence for the happiness and benefit of youth as may be found in any city in the land.

Six years later, on October 9, 1936, a *Standard* news item testifies to the success and loyal enthusiasm accorded the programs by the kids, themselves. The headline read, "Small Boys Protect Goal Posts at Parrot Flats." The news story:

> Two men using a small truck, attempted to dig up the goal posts on the girl's fieldball grounds at the Parrot Flats.
>
> A group of small boys started throwing stones and were planning a mass attack when the men ran to their truck and drove off. The boys have a cabin nearby and are sworn to protect their playing field.
>
> Now irate neighbors are going to back the boys. They'll call the police, apprehend such law-breakers if possible and help local playground directors prosecute.

As the writer said in *Copper Camp:* "It means something to have been a kid in Butte."

Track meets traditionally provide plenty of opportunity for participation sports among boy and girl students.

Marble champions Jack Penney, Don Knievel, Claire Butler, and Frank Williams.

Upper left, grade school marble players, May, 1926; center left, boys and girls grade school basketball teams, 1937; lower left, adult volunteer grade school coaches and officials, 1928; upper right, finalists in kite contest, May, 1926; center right, grade school girls roller skating competition; lower right, St. Joseph's grade school Class A football champions, 1937.

The skaters

RAMSEY 49½

BLAINE	19⅔	IMM. CONCEPTION		ST. ANN'S	1
EMERSON	14½	JEFFERSON		ST. JOHN'S	1½
FRANKLIN		LINCOLN	3	ST. JOSEPH'S	36
GARFIELD	29½/3	LONGFELLOW	4½	ST. LAWRENCE	
GRANT	14	McKINLEY		ST. MARY'S	
GREELEY		MADISON		ST. PATRICK'S	36½
HARRISON		MONROE	5	WASHINGTON	
HAWTHORN		SACRED HEART	2	WEBSTER	69
HOLY SAVIOR		SHERMAN	2½	WHITTIER	1½

Webster School
Track Team
Champions 1933
Sponsored by the
Butte Exchange Club

Smithers
Butte

Back Row
Katherine Bogle
Dave Bodsy...
    mgr.
Raymond Harris
Richard Gallant
Andy Braun
Jack Reinwand
    I.L.W. Champ.
Robert Judd
James Ballard
    asst. coach
W.J. Ballard
coach

Middle row
Ben Tyrand
Wm Haivy K.
Meldin Wilber
Chas McKenzie
Ed Judd
Chas Osburne
Wm Carroll

Front Row
Billie Opie
John Coller
J.T. Evans
Joe Devine
George Prlain
Philip Yovelich
    I.L.W. Champ
Tom Laird
Rodney McIntire

Longfellow School
Championship
Butte
Class A League

St. Johns

Champs Class B League HC 1926?

280

Kids ice skating at old Holland Rink

Grown-ups posing at old Holland Rink

POSTSCRIPT

Some pictures seem to avoid our major classifications of Mines, City, and People — yet they are so much a part of Butte that they insisted upon being included somewhere in the book. So we've given them a special postscript of their own.

Centerville backyards in the 1960s had not changed much over the years.

Cemeteries age, as do communities, and offer mute history in the names and dates on tombstones. This part of Mt. Moriah Cemetery, with "The Hill" in the background, is quietly symbolic of life, death and history of a great mining camp.

Anyone who remembers the baseball pools, two-bit tickets, and thousands to be won remembers Clifford's.

One of the most famous restaurant names across the nation, the old Chequamegon on North Main.

The Cornish pasty — "letter from 'ome" — is as traditional to Cousin Jack miners as the mines, themselves. A pasty is a man-filling cold meal in the "lunch bucket" or a magnificent dinner after work, served hot and liberally lashed with brown gravy. Customary ingredients: beef, potatoes, onions all diced "the size of the third joint of a Cornish woman's little finger." Add a bit of kidney suet, salt and pepper to taste, mix, fold pie crust over it all, crimp the edge, and bake in a hot oven. That's a pasty!

Sheriff Charles Henderson, reins in hand

The first airmail plane arrives in Butte - Aug 1 - 1928

Winter called for a graceful cutter and furs

When the Thornton Hotel ended its long history as a hotel, The Company converted it to a recreation center for employees.

1890

The Clark mansion

The last cabin in Highland City, the true ghost mining camp south of Butte

HIS HONOR, THE MAYOR . . .

These are some of the men who at one time or another served as the mayor of Butte —

Henry Jacobs 1879-80

Frank Beal 1881-82

O. B. Whitford 1883-84

William Owsley 1884-85

H. L. Frank 1885-86

W. R. Kenyon 1887-88

L. J. Hamilton 1888-89

H. G. Valiton 1890-91

Henry Mueller 1891-92

Lee Mantle 1892-93

E. O. Dugan 1893-95

William Thompson ... 1895-97

P. S. Harrington 1897-99

J. H. McCarthy 1899-01

William Davey 1901-03

Pat Mullins 1903-05

John MacGinniss 1905-07

Joseph Corby 1907-09

Charles Nevin 1909-11

Lewis J. Duncan 1911-15

Charles Lane 1915-17

W. H. Maloney 1917-19

W. T. Stodden 1919-21

James Cocking 1921-23

W. D. Horgan 1923-27

K. M. Beadle 1927-31

Archie McTaggart 1931-35

Charles Hauswirth ... 1935-41

Barry O'Leary 1941-49

Tom Morgan 1949-53

Tim J. Sullivan 1953-57

William Donnelly 1957-59

Vern D. (Hanna)
Griffith 1959-64

Tom Powers 1964-69

Mario Micone 1970-

289

C. Owen Smithers, Sr.

Epilogue:

To every thing there is a season, and a time to every purpose under the sun:

A time to be born, and a time to die; a time to plant, and a time to pluck up that which is planted;

A time to kill, and a time to heal; a time to break down, and a time to build up;

A time to weep, and a time to laugh; a time to mourn, and a time to dance;

A time to cast away stones, and a time to gather stones together; a time to embrace, and a time to refrain from embracing;

A time to get, and a time to lose; a time to keep, and a time to cast away;

A time to rend, and a time to sew; a time to keep silence, and a time to speak;

A time to love, and a time to hate; a time of war, and a time of peace.

Ecclesiastes, 3

INDEX